MINI-FARMING FOR A SELF-SUFFICIENT LIFESTYLE

A PRACTICAL GUIDE TO GROWING, PRESERVING, AND THRIVING ON A
SMALL PATCH OF LAND

HARPER MOSS

Feel free to join our mailing list for updates and information about other upcoming books by Harper Moss.

HARPER MOSS

Legal Notice

This book offers general information and inspiration about mini-farming and sustainable living. While every effort has been made to ensure the content is accurate and up-to-date, Harper Moss makes no guarantees—express or implied—regarding the accuracy, completeness, reliability, or suitability of the information provided.

Readers are encouraged to exercise discretion and seek qualified professional guidance when attempting projects, farming methods, or lifestyle changes described in these pages. The author and publisher accept no responsibility for any injuries, losses, or damages resulting from the use or misuse of the content.

Any product names, logos, or brand references mentioned are the property of their respective owners. These references are used solely for identification and informational purposes and do not imply endorsement.

Disclaimer

The insights and methods shared in this book are derived from the author's personal experience, research, and perspective. While we hope the ideas and advice prove valuable, we cannot guarantee results, as individual circumstances and outcomes may vary.

The practices and projects outlined in this book may not be suitable for all readers. Please consult appropriate experts or professionals before engaging in any activities that could pose a risk or require specialized knowledge.

By using this book, you acknowledge and agree that the author and publisher will not be held liable for any direct or indirect consequences arising from the use of the information contained herein. You assume full responsibility for your actions and decisions.

*C*ontents
Table of

DIY PROJECTS

INTRODUCTION

The Turn Toward Sustainable Food Production and Self-Sufficiency

It hit me one day—right there in the grocery store, staring at a bag of organic tomatoes with a price tag that made me shake my head. "This is ridiculous," I thought. "I could grow this myself." That passing frustration planted a seed in my mind that quickly grew into something bigger: a commitment to mini-farming and a more self-sufficient lifestyle.

With food shortages, skyrocketing prices, and unpredictable supply chains becoming the new normal, many of us are rethinking how we get our food. The reliance on industrial agriculture and big-box grocery stores is no longer just expensive—it's unsustainable. The idea of growing our own food, feeding our families, and living independently of that fragile system has re-emerged, echoing the values of our grandparents' generation.

Mini-farming is at the heart of that change. Whether you're working with a small backyard plot or a modest acre in the countryside, people everywhere are rediscovering the joy—and the power—of growing their own food. It's about more than just fresh produce; it's about control, security, and a deeper connection to what nourishes us.

And the best part? You don't need a giant farm or decades of experience. With a solid plan, some elbow grease, and a willingness to learn, anyone can start.

This movement toward self-sufficiency isn't just about saving money (although your wallet will notice). It's about peace of mind. It's about knowing what's in your food and where it came from. It's about sitting down at your dinner table and feeling the quiet pride of serving something you grew with your own two hands. And it's about choosing a healthier, more sustainable way of life that prioritizes people over profit.

What Makes Mini-Farming Different from Industrial Agriculture?

Say the word "farming," and most people picture massive operations—tractors rolling over endless fields of corn or wheat, rows of identical crops stretching to the horizon. But mini-farming couldn't be more different.

This is farming on a personal scale. It's intentional. It's a return to working with nature, rather than against it. While industrial agriculture focuses on mass production—often at the expense of nutrition, soil health, and environmental well-being—mini-farming emphasizes sustainability, biodiversity, and food grown for nourishment, not just profit.

Industrial agriculture turns food into a commodity. It depletes the land, relies on synthetic chemicals, and often produces crops stripped of their natural flavor and vitality. Mini-farming, on the other hand, creates an ecosystem. You compost, rotate your crops, collect rainwater, and use natural pest control. You find clever ways to make the most of your space—stacking functions, interplanting, and designing systems that support each other.

And the payoff? Food that's fresher, healthier, and packed with flavor—food that you can trust.

Now, let's be clear: mini-farming isn't magic. It's hard work. You'll face setbacks—pests, plant diseases, weather surprises. You'll experiment, and sometimes you'll fail. But then you'll bite into a tomato you grew yourself, bursting with real flavor, and it'll all make sense. Once you've tasted that difference, there's no going back.

Why Choose Mini-Farming?

So why start a mini-farm in the first place? Beyond the satisfaction of growing your own food, there are three big reasons: better nutrition, lower food costs, and greater self-reliance.

Healthier Food

Let's face it—supermarket produce often falls flat. It may look perfect, but taste and nutrition usually take a backseat. When you grow your own food, you control what goes into the soil and onto your plants. There's no long-distance shipping to drain nutrients, and no need for synthetic pesticides or fertilizers.

Homegrown food is simply better. It tastes fresher, it's more nutritious, and it connects you to the seasons. If you've ever plucked a warm tomato straight from the vine and eaten it in the garden, you know exactly what I mean.

Cost Savings

Trips to the grocery store aren't getting any cheaper—and if you're buying organic, the sticker shock is real. Mini-farming can significantly cut your food bills.

A well-planned garden can produce a large portion of your family's vegetables for a fraction of the cost. Add backyard chickens into the mix, and you're also saving money on eggs and meat. Even growing small batches of grains can reduce your reliance on store-bought staples. And when you learn to preserve your harvest—by canning, freezing, or fermenting—you stretch your savings throughout the year.

Self-Reliance

We've all seen how fragile the global food system really is. Whether it's a pandemic, supply chain delays, or extreme weather, disruptions can come out of nowhere. Mini-farming gives you a measure of control. You rely less on external systems and more on your own efforts.

But self-reliance isn't just about food—it's a mindset. It's about gaining skills, building confidence, and shifting your relationship with the land. Whether you're growing herbs on a balcony or raising chickens on a quarter-acre, you start to see your home not just as a place to live, but as a place to thrive.

What This Book Will Teach You

This book is your complete guide to starting and thriving with a mini-farm. Whether your goal is to grow some of your own food, become fully self-sufficient, or even turn your mini-farm into a small business, the chapters ahead will walk you through every step.

Here's what you'll find:

- ☑ **Chapter 1:** Foundations of Mini-Farming
- ☑ **Chapter 2:** Designing a Small-Scale Farm
- ☑ **Chapter 3:** Soil Composition and Management
- ☑ **Chapter 4:** Composting for Healthy Soil
- ☑ **Chapter 5:** Watering and Irrigation Systems
- ☑ **Chapter 6:** Understanding Plant Nutrients
- ☑ **Chapter 7:** Selecting Seeds and Starting Plants
- ☑ **Chapter 8:** Saving and Organizing Seeds
- ☑ **Chapter 9:** Crop Timing, Rotation, and Season Extension
- ☑ **Chapter 10:** Perennials, Fruit Trees, and Vines
- ☑ **Chapter 11:** Pest and Disease Control (The Integrated Approach)
- ☑ **Chapter 12:** Raising Chickens for Eggs
- ☑ **Chapter 13:** Raising Chickens for Meat
- ☑ **Chapter 14:** Butchering and Processing
- ☑ **Chapter 15:** Preserving Your Harvest – Canning, Freezing, Dehydrating
- ☑ **Chapter 16:** Other Preservation Techniques – Fermentation and Pickling
- ☑ **Chapter 17:** Processing Grains on a Small Scale
- ☑ **Chapter 18:** Marketing Your Products
- ☑ **Chapter 19:** Creating Value-Added Goods
- ☑ **Chapter 20:** Sustaining Self-Sufficiency for the Long Haul

Let's Get Growing

Whether you want to supplement your grocery list or build a life rooted in full self-sufficiency, this book will give you the tools and confidence to begin. So, roll up your sleeves, grab your gloves, and let's dig in. Your journey to a more self-sufficient lifestyle starts now.

PART 1:
PLANNING & SETTING UP YOUR MINI-FARM

CHAPTER 1: MINI-FARMING: THE FOUNDATIONS

Mini-farming isn't just a backyard hobby—it's a smart, sustainable way to grow your own food. It's about producing a surprisingly high yield on a small plot of land, all while nurturing soil health, conserving water, and minimizing your environmental impact. Unlike industrial agriculture—which focuses on high-output monocropping and heavy mechanization—mini-farming is designed for efficiency, self-reliance, and harmony with nature.

The Philosophy Behind Mini-Farming

At its core, mini-farming is a deliberate rejection of industrial agriculture's fast-paced, high-impact approach. It embraces organic practices, biodiversity, and a deep respect for the natural processes of growth and regeneration.

Where industrial agriculture relies on chemical fertilizers, genetically modified seeds, and monocultures, mini-farming emphasizes:

Healthy soil through composting, crop rotation, and organic amendments

Biodiversity by growing a range of vegetables, herbs, fruits, and even raising small livestock

Low environmental impact through reduced fossil fuel use, water conservation, and natural pest control

Self-sufficiency through independence from industrial food systems and supermarket chains

Mini-farming isn't just about food production—it's about reclaiming personal agency in how we eat and live.

Mini-Farming vs. Industrial Agriculture: A Side-by-Side Comparison

Aspect	Industrial Agriculture	Mini-Farming
Scale	Hundreds to thousands of acres	Small plots, backyards, rooftops, or urban spaces
Crop Diversity	Monoculture (single or limited crops)	Polyculture (crop rotation, intercropping, and variety)
Soil Health	Heavy use of synthetic fertilizers, deep tillage	Composting, organic soil amendments, low/no-till practices
Pest Control	Chemical pesticides and herbicides	Companion planting, natural predators, Integrated Pest Management (IPM)
Water Use	High-volume, often wasteful irrigation	Drip irrigation, rainwater harvesting, efficient watering
Labor	Mechanized, chemical-dependent	Hands-on, labor-intensive, rewarding
Sustainability	Depletes natural resources, pollutes ecosystems	Regenerates soil, protects biodiversity
Food Quality	Often lower in nutrients and flavor	Nutrient-dense, fresh, flavorful
Primary Goal	Maximize profit through yield	Maximize food security, sustainability, and independence

Why Mini-Farming Fits Today's World

The modern food system is vulnerable. From climate instability and economic uncertainty to global supply chain disruptions, we've seen just how easily access to food can be threatened. Mini-farming offers an empowering alternative: grow your own food, improve your health, and reduce your environmental footprint—all at the same time.

It's a response to rising prices, concerns about chemical-laden produce, and a growing desire to reconnect with nature. When you grow your own food, you don't just eat better—you know exactly what you're putting on your plate.

How Mini-Farming Works for Different Lifestyles

One of mini-farming's biggest strengths is its flexibility. You don't need acres of land or expensive equipment to get started. You can design a thriving mini-farm tailored to your space, budget, and schedule.

- **Urban Settings:** Use rooftops, balconies, or vertical gardening systems to grow produce in tight quarters.
- **Suburban Homes:** Transform backyards, front lawns, or community garden plots into productive spaces that include chickens, herbs, and vegetables.
- **Rural Properties:** With more room, expand into larger livestock, perennial crops, and even small-scale trade or sales.

Whether you have ten acres or ten feet, mini-farming can work for you.

A Shift in Mindset

Mini-farming isn't just a method—it's a perspective. It encourages you to see your land, however small, as a valuable resource. It asks you to think like a grower, to solve problems creatively, and to build skills over time.

You'll begin to see food not as something purchased in plastic packaging, but as something cultivated with care, rooted in patience and purpose.

Making the Most of Small Spaces

One of the biggest hurdles for new mini-farmers is working with limited space—but you'd be amazed at how much you can grow on a small plot when you use smart strategies.

Evaluating Your Growing Area

Start by assessing what you've got:

- **Sunlight:** Most food crops need 6–8 hours of full sun per day.
- **Soil:** Is it fertile? Will it need compost or amendments?
- **Structures:** Fences, walls, and trellises can support vertical growth.
- **Regulations:** Know your local zoning laws around water use, livestock, and urban agriculture.

Once you understand your space, you can design a layout that uses every inch efficiently.

Top Space-Saving Strategies

- **Vertical Gardening:** Use wall planters, trellises, and hanging baskets to grow upward.
- **Raised Beds:** Provide good drainage, improved soil control, and tight plant spacing.
- **Square Foot Gardening:** Divide your space into compact, productive blocks.
- **Intercropping:** Grow smaller plants like lettuce beneath taller ones like tomatoes.
- **Succession Planting:** Stagger planting times to harvest continuously throughout the season.
- **Container Gardening:** Great for patios, balconies, and driveways.
- **Hydroponics & Aquaponics:** Soil-free systems ideal for tight indoor or outdoor spaces.

Incorporating Small-Scale Livestock

Even in a compact space, you can raise animals that enhance your mini-farm:

- **Chickens:** Require minimal space, produce eggs, and help with composting.
- **Rabbits:** Provide lean meat, are quiet, and breed efficiently.
- **Quail:** A smaller alternative to chickens—great egg producers in tight quarters.
- **Bees:** Low-maintenance pollinators that also give you honey and wax.

Permaculture Principles in Mini-Farming

Permaculture is the art of designing agricultural systems that mimic natural ecosystems. It's an essential part of long-term success in mini-farming because it promotes sustainability, balance, and low-input productivity. Here's how permaculture enriches your mini-farm:

1. **Observe Your Land:** Study your space—sunlight, wind, water flow—before planting. Design with your environment, not against it.
2. **Capture and Store Energy:** Use rain barrels, mulch to retain moisture, and passive solar greenhouses.
3. **Get a Yield:** Every element should serve multiple functions—fruit trees can feed you, provide shade, and drop organic matter.
4. **Use Feedback Loops:** Monitor your system and adjust—change watering schedules, tweak soil, shift planting techniques.
5. **Prioritize Renewable Resources:** Use compost, mulch, and natural pest controls instead of synthetic chemicals.
6. **Eliminate Waste:** Turn food scraps into compost, reuse water, and repurpose materials.
7. **Encourage Diversity:** Grow multiple crops and raise varied animals for stronger resistance to pests and diseases.
8. **Integrate, Don't Separate:** Let the system work together—chickens control pests, trees shelter delicate plants, and flowers draw pollinators.

Mini-farming empowers you to feed yourself while respecting the land. It's more than a set of techniques—it's a lifestyle rooted in intention, adaptability, and sustainability.

In the chapters ahead, we'll dig into the practical side of building your mini-farm—starting with the soil beneath your feet.

CHAPTER 2: SMALL-SCALE FARM DESIGN

Designing your small-scale farm begins with thoughtful planning. How you organize your space has a direct impact on your food output, ease of management, and the overall sustainability of your efforts. Whether you're working with a spacious backyard, a compact urban lot, or even a patio or balcony, selecting the right layout—raised beds, row gardening, or container growing—can help you make the most of what you have.

Assessing Your Space and Needs

Before putting anything in the ground, take time to assess your space. Your design will be shaped by factors like:

- **Size of Your Growing Area:** Are you working with a balcony, rooftop, suburban backyard, or larger rural property?
- **Sunlight Exposure:** Identify areas with 6–8 hours of full sun daily. Match plant selection to light availability.
- **Soil Quality:** Poor or contaminated soil may mean you'll need raised beds or containers.
- **Water Access:** Proximity to a water source affects your irrigation setup.
- **Ease of Movement:** Paths should allow you to easily water, weed, harvest, and tend crops.

With these factors in mind, let's explore the three primary small-scale growing methods: raised beds, row gardening, and container farming.

Raised Beds: Efficient, High-Yield Farming

Raised beds are one of the most effective methods for small-scale farming. They concentrate nutrients, improve soil structure, and optimize drainage. They also enable you to grow more food per square foot than traditional methods.

Benefits of Raised Beds:

- **Improved Soil Quality:** You control what goes into the bed—ideal for managing nutrient content and avoiding contamination.

- **Higher Yields:** Intensive planting techniques result in more harvest per square foot.
- **Fewer Weeds:** Clear boundaries and dense planting reduce competition from weeds.
- **Better Drainage:** Especially helpful in areas prone to waterlogging.
- **Accessibility:** Beds can be built at waist height for those with mobility limitations.

Designing and Building Raised Beds:

- **Width:** Keep beds between 3–4 feet wide so you can reach the center without stepping in.
- **Depth:** Aim for 6–12 inches minimum; root crops may need up to 24 inches.
- **Materials:** Use treated wood, stone, metal, or brick—whatever fits your climate and budget.
- **Orientation:** Align beds to receive maximum sunlight, usually running north to south.
- **Soil Mix:** Blend compost, topsoil, and organic matter for a nutrient-rich medium.

Maximizing Raised Bed Efficiency:

- **Square-Foot Gardening:** Divide beds into 1x1 foot sections for efficient spacing.
- **Companion Planting:** Place mutually beneficial plants together to reduce pests and improve growth.
- **Vertical Growing:** Use trellises and supports to grow vining crops upwards and free up space.

Row Gardening: A Classic for Larger Plots

Row gardening remains a tried-and-true technique, especially when you have more ground to work with. Crops are planted in long rows with walking paths in between.

Advantages of Row Gardening:

- Ease of Maintenance: Rows are ideal if using small equipment for tilling, weeding, or harvesting.
- Good Airflow: Wide row spacing reduces the risk of fungal disease.
- Simplified Crop Rotation: Clear structure makes it easy to alternate crops annually.

Designing an Effective Row Garden:

- Orientation: Arrange rows from north to south to ensure even sun exposure.
- Spacing: Leave 12–24 inches between rows, depending on plant type and access needs.
- Irrigation: Drip irrigation and soaker hoses work especially well here.
- Intercropping: Pair slow- and fast-growing plants in the same row for maximum efficiency.

While row gardening is ideal for larger spaces, it may not be practical for compact urban gardens where every square foot counts.

Container Farming: Urban and Space-Smart

Container farming is highly adaptable and perfect for balconies, rooftops, patios, and areas with poor soil. With the right setup, nearly any crop can thrive in a container.

Benefits of Container Farming:

- **Flexible Placement:** Move containers as light and weather conditions change.
- **Pest and Weed Control:** Containers reduce issues with invasive pests and weeds.
- **Customized Soil:** Tailor the soil mix to each plant's needs.
- **Space-Saving:** Ideal for tight urban areas or apartment living.

Choosing the Right Containers:

- **Large Containers:** Tomatoes, peppers, and root crops do best in 5–10 gallon pots.
- **Small Containers:** Perfect for herbs, salad greens, and shallow-root vegetables.
- **Materials:** Options include plastic, ceramic, fabric grow bags, and wood—all with different drainage and insulation properties.
- **Drainage:** Ensure containers have adequate holes to prevent root rot.

Tips for Space Optimization:

- **Vertical Growing:** Use stacked pots, wall planters, and trellises.
- **Hydroponics and Aquaponics:** Soil-free systems ideal for compact areas.
- **Grouping:** Plant compatible crops together to maximize yields and reduce resource competition.

Combining Techniques for Maximum Output

Many mini-farmers mix methods to suit their space and goals. Here are some useful combinations:

Raised Beds + Containers: Use beds for main crops, containers for herbs and mobile plants.

Row Gardening + Raised Beds: Ideal for larger plots with rows for staples like corn and potatoes, and beds for greens and herbs.

Vertical Gardens + Containers: Perfect for small spaces—grow up and out at the same time.

Your ideal layout will depend on your available land, climate, water access, and food production goals. Flexibility is key.

Smart Strategies: Vertical Gardening, Companion Planting & Intercropping

In a small-scale farm, every square foot counts. These three techniques help you make the most of limited space:

Vertical Gardening:

Grow upward to save ground space:

- Trellises for beans, cucumbers, and tomatoes.
- Hanging planters for herbs and strawberries.
- Stackable beds or shelves for microgreens.
- Green towers and living walls for continuous harvests.

Companion Planting:

Pairing crops that help each other thrive:

- **Tomatoes + Basil:** Basil enhances flavor and repels pests.
- **Corn + Beans + Squash (Three Sisters):** Beans fix nitrogen, corn supports the beans, and squash shades the soil.
- **Carrots + Onions:** Onions repel carrot flies.

Intercropping:

Plant multiple crops in the same space for better yield and pest control:

- **Row Intercropping:** Alternate different crops in adjacent rows.
- **Strip Intercropping:** Use wider strips for larger crops and improved biodiversity.
- **Relay Intercropping:** Plant new crops before older ones are harvested to maintain constant productivity.

Encouraging Biodiversity and Natural Pest Control

Biodiversity strengthens your mini-farm's ecosystem and reduces dependence on synthetic chemicals.

Benefits:

- **Natural Pest Control:** Ladybugs, lacewings, and predatory wasps help manage harmful insects.
- **Improved Soil Health:** Diverse root systems enhance soil aeration and nutrient cycling.
- **Boosted Pollination:** Bees, butterflies, and other pollinators improve fruit and seed production.
- **Resilience:** A variety of crops prevents pests and diseases from spreading rapidly.

Methods:

- Plant pest-repellent species like marigolds, garlic, and basil near vulnerable crops.
- Attract beneficial insects with dill, fennel, and yarrow.
- Rotate crops annually to break pest cycles.
- Use mulch, netting, and row covers as barriers.
- Avoid chemical pesticides that disrupt natural ecosystems.

Planning for Self-Sufficiency: Crop Ratios and Food Security

Achieving self-sufficiency requires careful planning. Your mini-farm should meet your household's dietary needs while considering yield potential, storage, and seasonal variation.

Step 1: Understand Your Household's Needs

- **Calories:** Estimate daily and annual calorie requirements (2,000–2,500 per adult).
- **Nutrition:** Ensure a balance of carbs, proteins, fats, and micronutrients.
- **Taste Preferences:** Grow food your household enjoys.
- **Preservation:** Include crops that store well or can be canned, frozen, or dried.

Step 2: Plan Crop Categories

- **Carbs:** Potatoes, sweet potatoes, winter squash, grains.
- **Proteins:** Beans, lentils, peas, nuts.
- **Vegetables:** Leafy greens, carrots, beets, broccoli, etc.
- **Fruits:** Apples, berries, melons, citrus.
- **Herbs & Aromatics:** Garlic, onions, basil, thyme, oregano.

Step 3: Estimate Yield Per 100 Square Feet

- **Potatoes:** 100–150 lbs
- **Tomatoes:** 150–200 lbs
- **Carrots:** 30–50 lbs
- **Corn:** 10–20 ears
- **Dry Beans:** 10–20 lbs
- **Winter Squash:** 100–200 lbs

Use these numbers to calculate how much growing space you'll need per crop.

Extend the Season & Rotate Crops

- **Crop Rotation:** Rotate legumes, root vegetables, and heavy feeders to preserve soil fertility.
- **Season Extension:** Use greenhouses, cold frames, and row covers to grow earlier and longer.

Include Perennials for Long-Term Return

- **Perennials:** Fruit trees, berry bushes, rhubarb, asparagus.
- **Annuals:** Provide variety but need replanting each year.

Preserve Your Harvest

- **Canning:** Great for tomatoes, beans, and fruits.
- **Freezing:** Excellent for greens and many vegetables.
- **Drying:** Herbs, fruit, and certain vegetables.
- **Root Cellaring:** Store potatoes, onions, carrots, and squash.

Designing an efficient, productive small-scale farm is a balance of creativity, science, and practicality. Whether you're working with raised beds, containers, rows—or all three—the key is to match your method to your space, lifestyle, and goals. With good design and smart strategies like vertical gardening, biodiversity, and crop planning, your mini-farm can provide year-round abundance.

In the next chapter, we'll explore one of the most crucial foundations of successful mini-farming: understanding and caring for your soil.

CHAPTER 3: MANAGEMENT AND COMPOSITION OF SOIL

Understanding Soil Fertility

Every successful mini-farm begins with healthy soil. Without it, plants struggle to grow, become more vulnerable to pests and disease, and yield less. Soil fertility—determined by factors like pH balance, organic matter, and nutrient content—directly influences how productive and sustainable your garden or farm can be.

Fertile soil holds moisture well, offers a supportive structure for roots, and delivers the essential nutrients plants need to thrive. By understanding and managing these key components, you can ensure a vibrant, productive mini-farm season after season.

The Role of pH Balance in Soil Health

Soil pH measures its acidity or alkalinity on a scale from 0 to 14. Most crops prefer a slightly acidic to neutral range, between **pH 6.0 and 7.0**, where nutrients like nitrogen, phosphorus, and potassium are most accessible.

- **Acidic soil:** pH less than 7
- **Neutral soil:** pH 7
- **Alkaline soil:** pH above 7

When soil is too acidic or too alkaline, nutrients may become unavailable or even toxic, leading to stunted growth or poor yields.

How to Test and Adjust pH:

- Use a **home testing kit** or send samples to an agricultural extension lab.
- To **raise pH** (reduce acidity), add **agricultural lime** or **dolomitic lime.**
- To **lower pH** (reduce alkalinity), apply **sulfur, peat moss,** or **pine needles.**
- Retest regularly and adjust as needed to maintain optimal pH.

The Importance of Organic Matter

Organic matter—made up of decomposed plant and animal material—is one of the most critical components of fertile soil. It acts like a sponge, retaining water, loosening compacted areas, and slowly releasing nutrients. Ideal soil contains 3% to 6% organic matter.

Benefits of Organic Matter:

- **Moisture Retention:** Reduces the need for frequent watering.
- **Improved Soil Structure:** Promotes aeration and allows roots to spread freely.
- **Nutrient Supply:** Breaks down slowly to provide steady access to minerals.
- **Microbial Life:** Supports beneficial organisms that boost soil health and plant growth.

Ways to Build Organic Matter:

- Add **compost, aged manure, or mulched yard** waste regularly.
- Plant **cover crops** like clover or vetch to fix nitrogen and build biomass.
- Apply **organic mulch** such as straw, wood chips, or shredded leaves.
- Minimize tilling, which can degrade organic matter and disrupt beneficial microbes.

Essential Soil Nutrients: Macro and Micronutrients

Soil nutrients fall into two categories: **macronutrients**, required in large quantities, and **micronutrients**, needed in smaller amounts but still vital to plant health.

Primary Macronutrients:

- **Nitrogen (N)** – Encourages leafy growth; found in compost, manure, and legumes.
- **Phosphorus (P)** – Supports strong roots and flowering; found in bone meal and rock phosphate.
- **Potassium (K)** – Promotes resilience and disease resistance; found in kelp, greensand, and wood ash.

Secondary Macronutrients:

- **Calcium (Ca)** – Strengthens cell walls; prevents blossom end rot.
- **Magnesium (Mg)** – Vital for photosynthesis; found in dolomitic lime and Epsom salts.
- **Sulfur (S)** – Aids in protein production; available in gypsum and composted manure.

Micronutrients:

- **Iron (Fe)** – Essential for chlorophyll production.
- **Zinc (Zn)** – Supports growth and enzyme function.
- **Copper (Cu)** – Aids in photosynthesis and disease resistance.
- **Manganese (Mn)** – Helps with enzyme activity.
- **Boron (B)** – Important for seed and fruit development.

A well-balanced soil contains all of these nutrients in proper proportions. Imbalances can lead to poor yields or unhealthy plants.

Keeping Soil Nutrient-Rich

Maintaining fertile soil requires proactive care:

- **Crop Rotation:** Different crops use and return different nutrients. Rotation prevents depletion and reduces pests.
- **Natural Amendments:** Compost, worm castings, and organic fertilizers improve soil health without harming beneficial organisms.
- **Green Manures:** Plant and till in crops like rye or clover to replenish nutrients.
- **Avoid Overuse of Synthetics:** Chemical fertilizers can destroy helpful microbes and harm soil structure over time.

Soil Testing and Targeted Amendments

To understand what your soil needs, regular testing is essential. This gives insight into pH, nutrient content, and texture—allowing you to make informed adjustments.

Soil Testing Methods:

- **Home Test Kits:** Quick and inexpensive for measuring pH, nitrogen, phosphorus, and potassium.
- **Lab Testing:** Provides comprehensive data including organic matter levels and soil texture.
- **Soil Jar Test:** A DIY method where soil is mixed with water and allowed to settle, revealing sand, silt, and clay layers.
- **Biological Indicators:** A healthy soil teeming with earthworms is often rich in organic matter and nutrients.

Modifying Different Soil Types

- **Sandy Soil:** Fast-draining and low in nutrients. Improve by adding compost, mulch, and organic matter.
- **Clay Soil:** Retains water but compacts easily. Mix in sand, gypsum, and organic matter to improve aeration.
- **Loamy Soil:** Ideal texture with good drainage and fertility. Maintain with compost and cover cropping.
- **Saline Soil:** Common in arid regions. Apply organic material and flush with clean water to reduce salt buildup.

The key is to amend according to your soil's unique composition and continually adjust as you grow.

Microorganisms and the Living Soil

Beneath the surface, an invisible army of microbes plays a crucial role in soil health. Bacteria, fungi, protozoa, and nematodes all contribute to nutrient cycling, disease prevention, and plant vitality.

Why Soil Microbes Matter:

- **Nutrient Recycling:** Microbes break down organic matter and release plant-accessible nutrients.
- **Improved Structure:** Fungi and bacteria bind soil particles, enhancing aeration and water movement.
- **Disease Suppression:** Beneficial microbes outcompete harmful ones, reducing the need for pesticides.
- **Plant Stimulation:** Some bacteria stimulate root growth and improve nutrient uptake.
- **Carbon Storage:** Soil organisms help sequester carbon, supporting climate resilience.

Types of Soil Microbes and Their Benefits

1. Bacteria – Nature's Recyclers

- **Nitrogen Fixers (e.g., Rhizobia, Azotobacter):** Convert atmospheric nitrogen into usable forms.
- **Decomposers (Bacillus, Actinomycetes):** Break down plant material and complex compounds.
- **Disease Suppressors (Pseudomonas, B. subtilis):** Produce natural antibiotics and enzymes.

2. Fungi – The Soil Builders

- **Mycorrhizal Fungi:** Form symbiotic relationships with roots, boosting phosphorus uptake.
- **Saprophytic Fungi:** Decompose organic material.
- **Pathogenic Fungi:** Harmful types exist, but healthy soil keeps them in check.

3. Protozoa – The Nutrient Managers

- **Flagellates and Amoebas:** Feed on bacteria and release nitrogen.
- **Ciliates:** Regulate microbial activity in moist soils.

4. Nematodes – The Soil Regulators

- **Beneficial Types:** Feed on microbes, recycling nutrients.
- **Harmful Types:** Can attack plant roots—balance is key.

How to Encourage Soil Microbial Life

- **Minimize Tillage:** Disturbing the soil too often disrupts fungi and bacteria networks.
- **Add Organic Matter:** Feed the soil life with compost, manure, and cover crops.
- **Mulch Regularly:** Regulates temperature, retains moisture, and provides food for microbes.
- **Use Compost Teas and Inoculants:** Boost microbial diversity and support plant health.
- **Rotate Crops and Practice Polyculture:** Increases diversity above and below ground, keeping soil systems vibrant.

Healthy soil is the foundation of a thriving mini-farm. By understanding its structure, managing nutrients, and fostering beneficial organisms, you create a self-sustaining system that supports high yields, nutrient-rich produce, and long-term success.

In the next chapter, we'll explore the art and science of **composting**, one of the most effective ways to build and maintain fertile soil for years to come.

MINI-FARMING FOR A SELF-SUFFICIENT LIFESTYLE

CHAPTER 4: HEALTHY SOIL COMPOSTING

Composting is one of the most powerful tools in a mini-farmer's toolkit. It improves soil quality, recycles organic waste, and boosts sustainability—all without requiring synthetic inputs. Whether you're using traditional composting or vermiculture (worm composting), the goal remains the same: to break down organic materials into rich, nutrient-dense soil amendments. Understanding the differences and benefits of each approach will help you choose the best fit for your farm.

Traditional Composting

Conventional composting uses natural biological processes—led by microbes, insects, and fungi—to transform organic waste into dark, crumbly humus that enriches the soil.

How It Works

Effective composting depends on four core elements:

- **Carbon-rich "browns"** (leaves, straw, cardboard) and **nitrogen-rich "greens"** (food scraps, grass clippings, manure) must be balanced.

- **Moisture** should mimic a wrung-out sponge to support microbial activity.

- **Aeration** is crucial; turning the pile introduces oxygen and prevents foul odors.

- **Heat** generated during decomposition accelerates the process and destroys pathogens and weed seeds.

Building a Compost Pile

1. **Choose a site** near your garden—shady and well-drained is ideal.

2. **Layer materials**, alternating browns and greens.

3. **Keep the pile moist**, but not soggy.

4. **Turn regularly** (every 1–2 weeks) to keep decomposition active.

5. The pile should reach 130–160°F to maximize breakdown.

6. Within 2 to 6 months, you'll have rich humus ready to use.

Advantages of Traditional Composting

- Handles large volumes of yard and kitchen waste.

- Generates a high-quality soil amendment.

- Heat naturally sterilizes compost, killing most weed seeds and pathogens.

Drawbacks

- Requires space and consistent maintenance.

- Can attract pests if not managed properly.

- Takes longer than other methods like vermiculture.

Vermiculture (Worm Composting)

Vermiculture uses specific worms—typically **red wigglers**—to break down organic material. The result is worm castings, a highly nutritious compost ideal for boosting soil fertility.

How It Works

Red wigglers consume food scraps and paper waste, digesting them into castings that are rich in nutrients and beneficial microbes. Worms thrive in moist, dark, well-ventilated environments and are ideal for processing kitchen waste.

Setting Up a Worm Bin

1. Choose a ventilated bin made of plastic or wood.

2. Add bedding such as shredded newspaper, coconut coir, or leaves.

3. Introduce red wigglers.

4. Feed them small amounts of fruit and veggie scraps, coffee grounds, and shredded paper.

5. In 2 to 3 months, you'll have finished compost ready for use.

Advantages

- Faster than traditional composting.
- Requires less space—can be done indoors.
- Ideal for reducing kitchen waste.

Limitations

- Cannot process large volumes of waste quickly.
- Worms are sensitive to temperature and moisture.
- Requires attention to food balance and pH.

Choosing the Right Composting Method

- Traditional composting is ideal for larger properties with high volumes of yard waste.
- Vermiculture works well in urban or small-scale settings where space is limited, and kitchen waste is the primary input.
- Many mini-farmers successfully combine both methods—worms for kitchen scraps, piles for garden waste.

Designing a Composting System for Any Farm Size

Whether you're managing a balcony garden or a multi-acre homestead, composting systems can be scaled to fit your needs.

Small-Scale Composting for Backyard Gardens

1. **Compost Bins:**

Great for suburban or urban settings, these contained systems minimize odor and pest issues.

2. **Worm Bins:**

Perfect for indoor or balcony composting—great for kitchen scraps and nutrient-rich worm castings.

3. Compost Tumblers:

Rotating bins speed up decomposition and are perfect for tight outdoor spaces with minimal maintenance.

Medium-Scale Composting for Homesteads

1. Three-Bin Systems:

Separate bins for active compost, curing compost, and finished compost allow for continuous rotation and productivity.

2. Wire or Pallet Bins:

Affordable and easy to expand, these open-air bins offer good airflow and can handle more material.

3. Hybrid Systems:

Use compost piles for yard waste and worm bins for kitchen scraps. This dual approach improves compost quality and flexibility.

Large-Scale Composting for Mini-Farms

1. Windrow Composting:

Long rows of compostable material turned periodically—ideal for farms with lots of manure or crop waste.

2. Aerated Static Piles:

Use air pipes or blowers to supply oxygen, eliminating the need for manual turning.

3. Manure Management Integration:

On farms with livestock, composting manure reduces odor, prevents nutrient runoff, and creates rich soil amendments.

Common Composting Mistakes—and How to Fix Them

Even seasoned farmers encounter composting issues. Here are the most common problems and how to solve them:

Compost Doesn't Heat Up

Cause: Too dry, too much brown material, or lack of greens.

Fix: Add moist nitrogen-rich materials like manure or food scraps; turn the pile more frequently.

Bad Odor

Cause: Excess moisture or too many greens.

Fix: Add dry browns like leaves or cardboard and increase aeration.

Dry, Inactive Pile

Cause: Not enough greens or water.

Fix: Moisten lightly and add green materials.

Pest Issues (Flies, Rodents, Maggots)

Cause: Exposed food scraps or unbalanced materials.

Fix: Bury food scraps, avoid meat and dairy, use enclosed bins.

Slow Decomposition

Cause: Cold weather, poor aeration, or imbalance.

Fix: Turn pile more often, check material balance, insulate during winter.

Worm Death in Vermiculture Bins

Cause: Overfeeding, acidity, or extreme temperatures.

Fix: Feed sparingly, use crushed eggshells to buffer pH, and maintain temperatures between 55–77ºF.

With the right approach, composting becomes a powerful ally in your journey toward self-sufficiency. Whether you're enriching a backyard garden or managing waste on a larger mini-farm, a well-managed compost system improves soil fertility, reduces trash, and enhances long-term sustainability.

Up next, we'll dive deeper into irrigation and water management—another key factor in maintaining a thriving, resilient growing system.

CHAPTER 5: IRRIGATION AND WATERINGS

Water is one of the most essential components of any farming operation. Without the right watering strategy, plants become stressed, yields drop, and soil quality deteriorates. For small-scale farmers striving for self-sufficiency, efficient irrigation is key to maintaining healthy crops while conserving this precious resource.

In this chapter, we'll explore smart irrigation systems, water-saving techniques, and optimal watering schedules to help you nourish your plants effectively and sustainably.

Efficient Irrigation Systems: Drip Irrigation, Soaker Hoses, and Rainwater Harvesting

Traditional irrigation systems—like overhead sprinklers—are often wasteful due to evaporation and runoff. Mini-farmers can benefit significantly from adopting efficient, targeted watering methods designed for small plots and high-impact results.

Drip Irrigation

Drip irrigation is one of the most efficient watering systems available. It delivers water directly to the base of each plant through a network of tubes and emitters, minimizing loss from evaporation and runoff.

Benefits of Drip Irrigation:

- Uses up to 50% less water than conventional methods.
- Keeps non-planted areas dry, which helps reduce weed growth.
- Prevents soil erosion and nutrient leaching.
- Can be automated with timers for consistent watering.

How to Set Up a Drip System:

- Lay tubing around plants or along crop rows.
- Place emitters near the base of each plant, adjusting flow for water needs.
- Connect the system to a hose, faucet, or rainwater collection barrel.
- Use a filter and pressure regulator to maintain flow and prevent clogs.
- Set a timer for scheduled efficient watering.

Soaker Hoses

Soaker hoses are porous tubes that release water slowly along their length. They're ideal for raised beds, row crops, and compact garden plots.

Benefits of Soaker Hoses:

- Even water distribution with minimal evaporation.
- Simple to install and relocate.
- Ideal for low-maintenance watering.

Rainwater Harvesting

Rainwater collection is a sustainable way to supplement your irrigation needs. By capturing runoff from rooftops, you reduce your dependence on municipal or well water and store water for dry periods.

How to Build a Rainwater System:

- Install gutters and downspouts on rooftops.
- Collect water in barrels or large storage tanks.
- Filter debris using mesh screens or first-flush diverters.
- Connect to your drip or soaker hose system using gravity or a pump.

Water-Saving Strategies and Drought Resilience

Whether you're in a drought-prone area or simply want to conserve, smart water-saving techniques make your mini-farm more efficient and environmentally responsible.

Mulching

Applying organic mulch (like straw, shredded leaves, or wood chips) around your plants:

- Retains soil moisture
- Regulates soil temperature
- Suppresses weeds
- Adds nutrients as it decomposes over time

Companion Planting

Strategically pairing plants that share similar moisture needs helps you use water more efficiently. Taller plants can also create shade and reduce evaporation for lower-growing companions.

Swales and Contour Farming

By shaping the land with shallow ditches (swales) along natural contours, you can slow runoff and direct water toward plant roots.

This not only improves water absorption but also helps prevent erosion.

Greywater Recycling

Lightly used household water from sinks, showers, and washing machines (known as greywater) can be filtered and redirected for irrigation—especially for non-edible plants or ornamental beds. Basic filtration and responsible use make this a practical option for water-conscious farmers.

Deep Watering Techniques

Instead of frequent shallow watering, aim to water deep and less often. This encourages roots to grow deeper, making plants more drought-tolerant and resilient in tough conditions.

Optimizing Watering Schedules for Plant Health

Getting the timing and frequency right is just as important as the method. Overwatering can lead to root rot and fungal disease, while underwatering can weaken plants and reduce yields.

Key Factors in Watering Frequency:

Soil Type:

Sandy soil drains quickly and needs more frequent watering. Clay soil retains water longer and needs less frequent watering.

Plant Maturity:

Seedlings need consistent moisture. Mature plants can tolerate longer dry spells.

Weather:

Adjust schedules during heat waves or after rainfall to avoid overwatering.

Time of Day:

Water early in the morning or in the evening to reduce evaporation and give plants time to absorb moisture.

Example Seasonal Watering Schedule:

Early Spring: Light, consistent watering to encourage seed germination.

Growing Season: Increase frequency as temperatures rise; maintain even soil moisture.

Harvest Season: Slightly reduce watering to concentrate fruit flavor and prevent mold.

Winter: Cut back watering as most plants go dormant—perennials still need occasional hydration.

Building a Resilient, Water-Efficient Farm

Efficient watering isn't just about saving time or money—it's about building a system that can adapt to environmental challenges. Drip irrigation, soaker hoses, and rainwater collection form the backbone of a smart irrigation setup. Add in mulching, companion planting, and deep watering strategies, and your farm becomes not only more productive but more drought-resistant and sustainable.

By understanding your soil, adjusting your watering schedule to the season, and investing in efficient systems, you give your plants the best chance to thrive—no matter the conditions.

In the next chapter, we'll dig into the nutritional needs of plants, exploring how macro and micronutrients work, how to identify deficiencies, and how to create a balanced nutrient program tailored to your mini-farm.

PART 2:
GROWING AND HARVESTING YOUR CROPS

CHAPTER 6: COMPREHENSIVE PLANT NUTRIENTS

Just like all living things, plants rely on a balanced and nourishing diet to grow strong, resilient, and productive. The nutrients they absorb directly influence root strength, leaf growth, fruit production, and overall plant health. For mini-farmers, understanding how macronutrients, micronutrients, and organic amendments support this development is essential for cultivating a thriving, self-sufficient garden.

In this chapter, we'll break down each type of nutrient, its purpose, how to recognize deficiencies, and natural ways to maintain soil richness and fertility.

Essential Nutrients for Plant Growth

Plants require a variety of nutrients categorized into **macronutrients** and **micronutrients**. Each plays a specific role, and a deficiency in even one can disrupt growth, reduce yields, and weaken resistance to pests or diseases.

Macronutrients: The Big Three (N-P-K)

The three major macronutrients are:

- Nitrogen (N)
- Phosphorus (P)
- Potassium (K)

These are present in the largest quantities in plants and are most often found in commercial fertilizers.

Nitrogen (N): For Lush Green Growth

Nitrogen is essential for the formation of chlorophyll—the compound plants use for photosynthesis—and supports vigorous leafy growth.

Functions:

- Stimulates rapid vegetative growth
- Enhances leaf color and overall vitality
- Supports protein and enzyme synthesis

Deficiency Signs:

- Yellowing of older leaves
- Pale green plants
- Slow or stunted growth

Natural Sources:

- Compost, aged manure, blood meal, fish emulsion, cover crops (e.g., clover, alfalfa)
- Synthetic: Urea, ammonium nitrate, ammonium sulfate

Phosphorus (P): For Roots, Flowers, and Fruits

Phosphorus helps in root development, flowering, fruiting, and energy transfer within the plant.

Functions:

- Strengthens root systems
- Boosts blooming and fruit formation
- Enhances photosynthesis and energy flow

Deficiency Signs:

- Dark green or purplish leaves, especially in young plants
- Weak or delayed flowering and fruiting

Natural Sources:

- Rock phosphate, fish meal, bone meal, compost
- Synthetic: Superphosphate, monoammonium phosphate

Potassium (K): For Strength and Resistance

Potassium helps regulate water movement, enzyme activity, and nutrient transport, building a plant's tolerance to stress and disease.

Functions:

- Improves drought and disease resistance
- Enhances fruit quality and flavor
- Strengthens stems and root systems

Deficiency Signs:

- Brown or scorched leaf edges
- Weak stems
- Poor fruit development

Natural Sources:

- Kelp meal, greensand, wood ash, composted banana peels
- Synthetic: Potassium sulfate, muriate of potash

Secondary Macronutrients: Calcium, Magnesium, and Sulfur

Though required in smaller quantities than N-P-K, these nutrients are vital to plant function and structure.

Calcium (Ca)

Supports root development and strengthens cell walls.

Deficiency: Blossom end rot in tomatoes and peppers, distorted new leaves

Sources: Crushed eggshells, gypsum, agricultural lime

Magnesium (Mg)

A key part of chlorophyll and photosynthesis.

Deficiency: Yellowing between leaf veins (interveinal chlorosis)

Sources: Dolomitic lime, Epsom salt

Sulfur (S)

Aids in enzyme formation and protein synthesis.

Deficiency: Pale leaves and stunted growth

Sources: Composted manure, gypsum, elemental sulfur

Micronutrients: The Unsung Heroes

Even though needed in trace amounts, micronutrients play vital roles in plant health. Deficiencies can lead to subtle but serious growth issues.

Micronutrient	Role	Deficiency Signs	Natural Sources
Iron (Fe)	Chlorophyll production	Yellowing between veins of young leaves	Compost, chelated iron
Zinc (Zn)	Hormone and enzyme function	Stunted growth, small leaves	Composted manure, zinc sulfate
Copper (Cu)	Reproduction, photosynthesis	Leaf curling, dieback	Compost, copper sulfate
Manganese (Mn)	Enzyme activation	Yellowing in young leaves	Compost, manganese sulfate
Boron (B)	Flower and fruit development	Brittle, distorted leaves	Borax (used sparingly), compost
Molybdenum (Mo)	Nitrogen fixation	Yellowing, curled leaves in legumes	Compost, molybdenum fertilizers

Organic Soil Amendments for Long-Term Fertility

To maintain healthy soil and plant nutrition, mini-farmers should prioritize organic materials that feed both the plants and the soil itself.

Compost

An all-purpose amendment that improves soil structure, encourages microbial life, and provides a wide range of nutrients. Apply regularly for best results.

Aged Manure

A rich source of nitrogen and organic matter. Always compost thoroughly before use to avoid plant burn and reduce pathogens.

Cover Crops

Plants like clover and alfalfa fix nitrogen in the soil, prevent erosion, and improve soil texture when turned in as green manure.

Mulch

Organic mulches (like straw or wood chips) help retain moisture, suppress weeds, and gradually decompose to enrich the soil.

Liquid Fertilizers

Compost tea, seaweed extract, and fish emulsion offer fast-acting nutrient boosts and improve soil microbial diversity.

Organic Fertilizers in Action: Compost Tea, Manure, and Biochar

Natural, homemade fertilizers enhance soil health without the negative side effects of synthetics. Three key options for small farms include:

Compost Tea

A liquid fertilizer made by steeping mature compost in water—packed with nutrients and beneficial microbes.

How to Make It:

1. Place compost in a mesh bag and submerge in a bucket of non-chlorinated water.

2. Aerate using a pump or stir occasionally for 24–48 hours.

3. Strain and apply directly to the soil or as a foliar spray every 1–2 weeks.

Benefits:

- Boosts microbial life
- Provides readily available nutrients
- Enhances disease resistance

Manure Applications

Different manures provide various benefits. All should be composted or aged for safety.

Manure Type	Benefits
Cow	Balanced nutrients; safe when aged
Chicken	High nitrogen; compost before use
Rabbit	Dry and nitrogen-rich; safe to use directly

Tips:

- Apply raw manure in fall or winter to allow breakdown before planting.
- Create manure tea by soaking aged manure in water for use as liquid fertilizer.

Biochar

A charcoal-like substance made from burning organic material in low oxygen (pyrolysis). Enhances soil structure and retains nutrients.

How to Make and Use:

1. Use hardwoods, crop residue, or coconut shells.

2. Burn in a controlled low-oxygen chamber (pit or kiln).

3. Crush and "charge" biochar with compost or manure before applying.

Benefits:

- Increases soil aeration and microbial life
- Reduces nutrient leaching
- Enhances long-term carbon storage

Diagnosing and Correcting Nutrient Deficiencies

Early recognition of nutrient deficiencies helps prevent crop failure and restores balance before damage becomes severe.

Common Deficiencies and Fixes:

Deficiency	Symptoms	Solutions
Nitrogen	Yellow older leaves, weak stems	Add aged manure, blood meal, compost tea
Phosphorus	Purple/blue leaves, weak roots	Use bone meal, rock phosphate
Potassium	Leaf edge browning, poor fruit	Apply greensand, kelp meal, wood ash
Magnesium	Yellowing between veins	Use Epsom salt, dolomitic lime
Calcium	Blossom end rot, curled new leaves	Add gypsum, eggshells, lime
Iron	Yellowing in new leaves	Apply chelated iron or organic compost
Zinc	Small leaves, poor fruit set	Add composted manure, zinc sulfate
Boron	Hollow stems, distorted growth	Use borax (sparingly), organic compost

Preventive Practices for Balanced Nutrition

Soil Testing: Use test kits or lab analysis annually to monitor nutrient levels.

Crop Rotation: Prevent nutrient depletion by alternating plant families.

Organic Amendments: Compost, manure, and biochar support long-term fertility.

Mulching: Reduces evaporation and nutrient leaching while slowly enriching the soil.

Nourishing the Soil, Feeding the Farm

Healthy plants come from healthy soil. When you focus on building soil fertility through organic practices and targeted nutrition, your mini-farm becomes more resilient, productive, and sustainable. With careful observation and well-informed adjustments, you'll maintain a thriving garden that supports both your crops and the ecosystem beneath the surface.

In the next chapter, we'll explore how to select seeds and start plants, guiding you through choosing resilient varieties and germination methods for your farm's unique needs.

CHAPTER 7: CHOICE AND STARTING OF SEEDS

The success of your small farm begins with one critical decision: selecting the right seeds and knowing how to start them. Whether you're sowing seeds indoors or directly into garden beds, understanding the differences between these methods—and following best practices—gives your crops their best chance to thrive.

Indoor Seed Starting vs. Direct Sowing

Some plants grow best when started indoors and transplanted later; others prefer to be sown directly where they'll mature. Factors such as plant variety, local climate, growing season length, and available resources all play a role in this decision.

Starting Seeds Indoors

Starting seeds indoors extends the growing season and gives you greater control over temperature, moisture, and lighting. This method is particularly useful in colder regions or where the outdoor season is short or unpredictable.

Benefits:

Extended Growing Time: Ideal for short seasons or late frosts.

Stronger Seedlings: Plants develop in a stable environment before facing outdoor stressors.

Protection from Pests and Disease: Limits early exposure to garden threats.

Controlled Conditions: Consistent light, warmth, and moisture boost germination success.

Best Crops to Start Indoors:

- Tomatoes
- Peppers
- Eggplants
- Broccoli
- Cauliflower
- Cabbage
- Herbs like basil, parsley, and thyme

How to Start Seeds Indoors: A Step-by-Step Guide

1. **Choose Containers:** Use seed trays, peat pots, or small cups with drainage holes.

2. **Use a Seed-Starting Mix:** Avoid using garden soil. Opt for a light, sterile, and well-draining mix.

3. **Plant at Proper Depth:** Follow seed packet instructions. Larger seeds go deeper; smaller ones stay near the surface.

4. **Provide Ample Light:** Seedlings need 12–16 hours of bright light daily. Use grow lights if natural light is inadequate.

5. **Maintain Moisture:** Keep the soil evenly moist—not soggy. Mist with a spray bottle as needed.

6. **Keep Warm:** Ideal germination temperature is between 65–75°F (18–24°C). A heat mat can help.

7. **Harden Off Before Transplanting:** Gradually introduce seedlings to outdoor conditions over 7–10 days to reduce transplant shock.

Direct Sowing

Direct sowing involves planting seeds straight into the soil where they will grow to maturity. Some crops prefer this method and do not transplant well.

Benefits:

Less Handling: Eliminates the transplanting step, reducing root disturbance.

Natural Root Development: Plants establish themselves where they'll remain.

Ideal for Root Crops: Vegetables like carrots and radishes prefer undisturbed soil.

Best Crops for Direct Sowing:

- Carrots
- Beets
- Radishes
- Turnips
- Lettuce
- Spinach
- Kale
- Arugula
- Peas
- Beans
- Squash
- Cucumbers
- Sunflowers

How to Direct Sow Successfully:

1. **Prepare the Soil:** Remove rocks and debris. Loosen soil and amend with compost.

2. **Check Soil Temperature:** Cold soil can delay or prevent germination. Wait for optimal warmth.

3. **Plant at Correct Depth and Spacing:** Refer to the seed packet. Typically, plant seeds at twice their width.

4. **Water Gently:** Keep the soil moist, but avoid overwatering, which can rot seeds.

5. **Thin Seedlings:** After germination, thin to the strongest plants for proper spacing and airflow.

Indoor or Direct? How to Choose

Your choice depends on your climate, frost dates, and farm setup. For example, starting cucumbers indoors may offer a jumpstart in cooler areas, while in warmer climates, they can be direct sown successfully.

Common Seed-Starting Mistakes to Avoid

1. Starting Too Early Indoors

Seedlings may outgrow their containers or become leggy if started too far ahead of the last frost.

2. Incorrect Planting Depth

Seeds planted too deep may not germinate; those planted too shallow may dry out.

3. Overwatering or Underwatering

Waterlogged soil leads to root rot; dry soil prevents germination. Keep the soil consistently moist, not soaked.

4. Skipping the Hardening-Off Process

Transferring indoor plants directly outside without acclimation often results in shock or death.

5. Poor Soil Preparation for Direct Sowing

Hard, compacted, or nutrient-poor soil reduces germination and weakens seedlings. Always enrich and loosen the soil before planting.

Choosing the Right Seeds for Your Soil and Climate

Success starts with the seed. Not all varieties thrive in every region or soil type. By selecting high-quality, climate-appropriate seeds, you'll set your farm up for strong yields, fewer pest problems, and healthier plants.

Know Your Climate

- **USDA Hardiness Zones:** Help determine which plants can survive winter temperatures in your area.
- **First and Last Frost Dates:** Essential for timing seed starting indoors or outside.
- **Growing Degree Days (GDDs):** Measure accumulated heat over time—critical for heat-sensitive crops.
- **Rainfall Patterns:** Some crops need more water, while others tolerate drought. Choose accordingly.

Understand Your Soil

Conduct a soil test to evaluate:

- **Soil Type:** Sandy, clay, or loamy soil affects drainage and nutrient retention.
- **pH Level:** Most crops prefer 6.0–7.0. Acid-loving plants (like blueberries) require lower pH.
- **Nutrient Content:** Deficiencies in nitrogen, phosphorus, or potassium can limit growth. Amend as needed before planting.

MINI-FARMING FOR A SELF-SUFFICIENT LIFESTYLE

Types of Seeds

- **Heirloom (Open-Pollinated):** Known for rich flavor and genetic diversity. Seeds can be saved and replanted.
- **Hybrids (F1):** Bred for consistency and disease resistance, but seeds aren't viable for saving.
- **Organic Seeds:** Grown without synthetic chemicals—ideal for eco-conscious farming.
- **Regionally Adapted Seeds:** Bred for local soil, pests, and climate. Often more resilient and reliable.

Seed Starting Mistakes and How to Avoid Them

Low-Quality or Old Seeds

Seeds lose viability over time. Always check expiration dates and conduct a germination test if in doubt.

Poor Soil or Medium

Dense or contaminated soil can smother seeds. Use a light, sterile, seed-starting mix for best results.

Inconsistent Moisture

Soil that's too wet may cause damping-off, a fungal disease. Soil that's too dry prevents germination. Keep moisture even and moderate.

Insufficient Light

Too little light leads to leggy, weak seedlings. Use grow lights or ensure 12–16 hours of bright natural light per day.

Skipping Hardening Off

Gradual exposure to outdoor conditions prevents transplant shock. Increase outdoor time each day over 7–10 days before planting out.

Growing from Seed: A Foundation for Farm Success

Understanding how and when to start seeds gives you control over your growing season and crop success. Whether you're leveraging the extended season from indoor starts or embracing the simplicity of direct sowing, attention to detail makes all the difference.

Pairing seed selection with climate awareness and soil preparation ensures a farm that is more resilient, productive, and sustainable.

In Chapter 8, we'll explore the art of seed saving and storage, equipping you with the knowledge to become more self-sufficient and carry your farm from one season to the next.

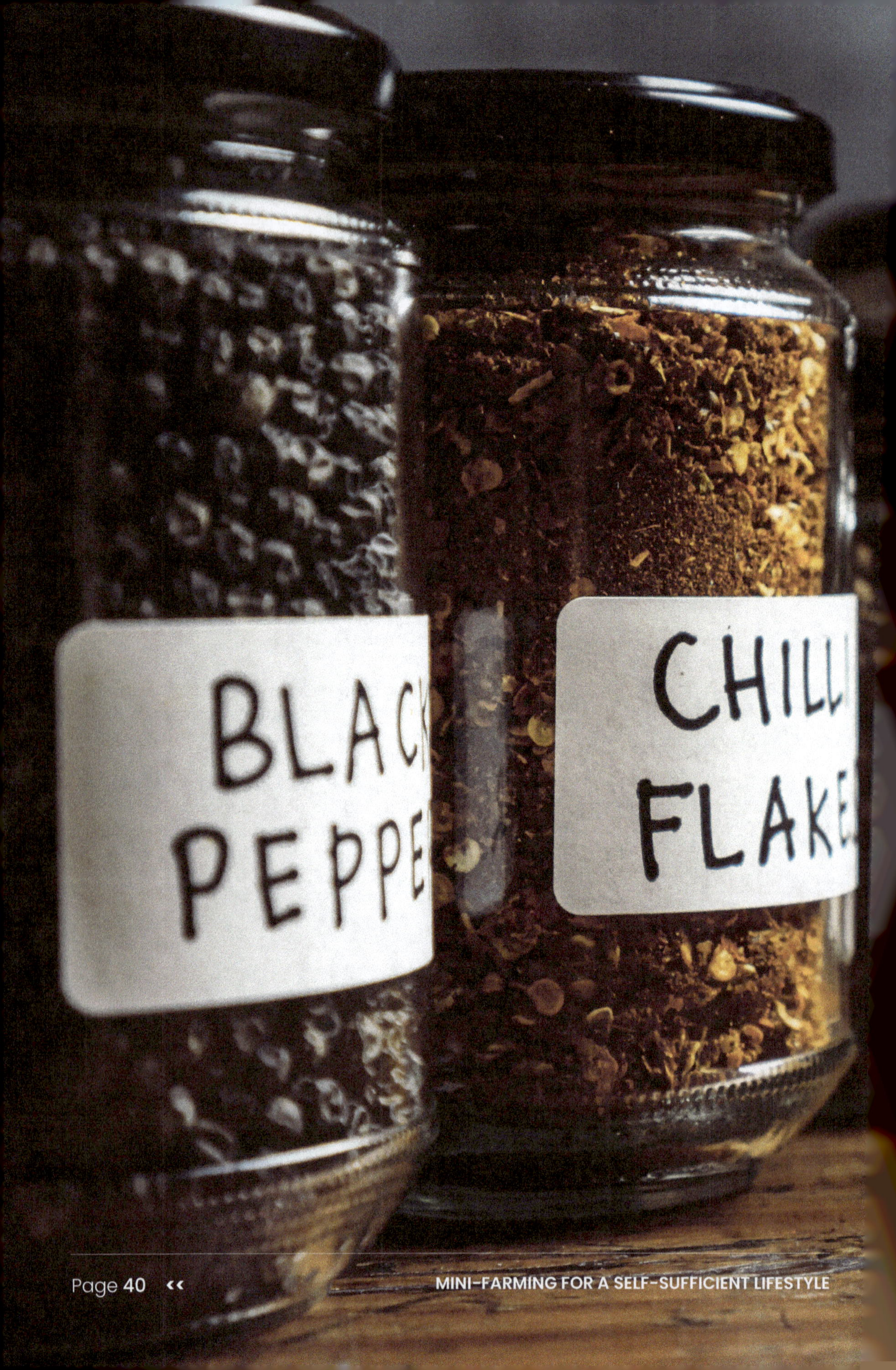
BLAC
PEPPE
CHILL
FLAKE

CHAPTER 8: CONSERVING AND ORGANIZING SEEDS

The Value of Heirloom Seeds for Sustainability

In the pursuit of sustainable farming and self-sufficiency, heirloom seeds play a central role. Unlike hybrid or genetically modified varieties, heirlooms have been carefully passed down through generations, preserving their original traits. These seeds offer exceptional advantages in terms of biodiversity, adaptability, cultural heritage, and long-term food security—making them a valuable asset for every small farm.

What Are Heirloom Seeds?

Heirloom seeds are open-pollinated varieties that have been cultivated and preserved for at least 50 years. These seeds are genetically stable and will reproduce true to type, which means they consistently produce plants with the same characteristics each generation. Often treasured by families or communities, heirloom seeds represent more than just food—they are a link to tradition, resilience, and agricultural knowledge.

Why Heirloom Seeds Matter for Sustainability

Genetic Diversity: Heirloom seeds preserve a broad gene pool, providing greater resilience against pests, diseases, and changing environmental conditions.

Local Adaptation: Many heirlooms have evolved over decades to suit specific climates, soils, and regional pest pressures. This makes them especially well-suited for small-scale, localized farms.

Superior Taste and Nutrition: Heirloom crops are renowned for their flavor, color, and nutritional value—often surpassing the appearance-focused hybrids bred for commercial markets.

Seed Sovereignty: Saving your own heirloom seeds reduces dependence on commercial seed companies and protects your farm's autonomy.

Cultural and Historical Preservation: Many heirloom varieties are tied to regional traditions and culinary heritage, helping to safeguard the stories of past generations.

Challenges with Heirloom Seeds

While heirlooms offer significant benefits, they also come with a few challenges:

Disease Susceptibility: Some heirloom varieties lack the disease resistance of modern hybrids. However, natural practices like crop rotation, companion planting, and organic pest control can mitigate this.

Variable Yields: Unlike hybrids bred for uniformity, heirloom crops can vary in productivity. Selecting seeds from your strongest plants helps improve consistency over time.

Storage Sensitivity: To remain viable, heirloom seeds must be stored properly in cool, dry, and dark environments.

Incorporating Heirloom Seeds on Your Mini-Farm

Start Small: Try a few heirloom varieties first to test how they perform in your climate and soil.

Learn Seed-Saving Techniques: Preserving viable seeds each season ensures sustainability.

Connect with Exchanges: Join local or online heirloom seed swaps to diversify your crops and connect with fellow growers.

Track Performance: Keep a gardening journal to monitor how each heirloom performs and adjust your selections each year.

By incorporating heirloom seeds into your farm, you help preserve traditional agriculture while building a more resilient and independent food system.

Seed Saving Guide for Different Plant Types

Saving seeds is an empowering and cost-effective way to build long-term self-sufficiency. Different types of plants require different techniques to ensure that the seeds remain viable and true to type.

Understanding Pollination Types

Seeds can be grouped based on how they reproduce:

Self-Pollinating: Plants fertilize themselves and produce genetically stable seeds. (Examples: beans, peas, tomatoes)

Cross-Pollinating: Require pollen from other plants of the same species, introducing genetic variation. (Examples: corn, squash, cucumbers)

Biennials: These plants produce seeds in their second growing season. (Examples: onions, carrots, beets)

Knowing the type of pollination is essential for maintaining seed purity.

Step-by-Step Guide: Saving Seeds by Crop Type

Tomatoes (Self-Pollinating)

Harvest: Choose fully ripe tomatoes from your healthiest plants.

Ferment: Scoop the seeds with pulp into a jar. Let it ferment for 2–3 days, stirring daily. This breaks down the gelatinous coating and kills harmful pathogens.

Clean: Add water, stir, and let the viable seeds settle. Pour off floating debris and rinse the seeds.

Dry: Spread seeds on a mesh screen or paper towel in a well-ventilated area. Dry thoroughly before storing.

Peppers (Self-Pollinating)

Harvest: Use mature peppers with full color.

Extract: Slice open and remove the seeds.

Dry: Spread seeds in a single layer in a warm, shaded space for several days.

Beans and Peas (Self-Pollinating)

Harvest: Let pods dry on the plant until they turn brown and brittle.

Extract: Shell the dry pods and collect seeds.

Dry: Air-dry seeds completely before storing.

Squash and Pumpkins (Cross-Pollinating)

Harvest: Select fully ripe fruits with hardened rinds.

Clean: Scoop seeds, remove pulp, and rinse thoroughly.

Dry: Spread in a single layer for 5–7 days in a ventilated space.

Corn (Cross-Pollinating)

Harvest: Leave ears on the stalk until kernels are hard.

Extract: Rub dry kernels from cobs.

Dry: Spread on screens to ensure full drying.

Onions, Beets, and Carrots (Biennials)

Overwinter: Leave plants in the ground or replant after storage.

Seed Harvest: In the second season, collect seed heads once they brown and dry.

Dry: Gently crush to release seeds and dry fully before storing.

Proper Seed Storage for Long-Term Viability

Storing seeds correctly ensures they remain viable for future planting seasons.

1. Ensure Seeds Are Fully Dry

Moisture is the enemy of stored seeds. To test dryness, try snapping a seed in half—if it bends, it needs more drying time.

2. Use Airtight Containers

Store seeds in:

- Glass jars with tight lids
- Mylar bags with desiccant packets
- Vacuum-sealed bags for maximum protection

3. Store in a Cool, Dark Place

Ideal storage temperatures range from 32°F to 50°F. A refrigerator, cold cellar, or insulated cabinet can provide stable conditions.

4. Label Everything Clearly

Include:

- Plant name and variety
- Harvest date
- Any growing notes or performance insights

5. Test Viability Annually

Place 10 seeds on a damp paper towel in a sealed plastic bag. If fewer than 7 sprout (70% germination), it's time to refresh your seed stock.

Building Seed Sovereignty on the Mini-Farm

By saving and storing seeds, mini-farmers take control of their food production, reduce long-term costs, and preserve the genetic legacy of plants well-suited to their region. This practice not only safeguards future harvests but also contributes to a deeper connection with the natural rhythm of farming.

In the next chapter, we'll explore how to time your planting, extend your growing season, and rotate crops to keep your soil healthy and productive year after year.

CHAPTER 9: TIMING, SEASON EXTENSION, AND CROP ROTATION

Building Resilience and Productivity Through Smart Growing Cycles

The Science Behind Crop Rotation

On a mini-farm, crop rotation is one of the most effective and time-tested methods for maintaining soil health and maximizing yields. By methodically rotating different crops in a planned sequence, farmers can prevent soil depletion, reduce pest populations, and improve overall soil fertility. For anyone aiming to create a thriving and sustainable growing system, understanding the science behind crop rotation is essential.

Why Crop Rotation Matters

Rotating crops follows a schedule that ensures different plant families are cultivated in a strategic sequence. The benefits are numerous:

Balanced Nutrient Use: Different crops draw from and contribute to the soil in unique ways. While some plants are heavy feeders, others restore essential nutrients. Rotation balances nutrient extraction and replenishment over time.

Natural Pest and Disease Control: Many pests and diseases are crop-specific. Changing the types of crops grown in each plot helps break pest and disease cycles naturally.

Improved Soil Structure: Varying root systems interact with the soil differently, enhancing aeration, preventing compaction, and increasing organic matter.

Boosted Soil Microbiology: A diverse range of crops fosters beneficial microbial activity, enriching the overall soil ecosystem.

Managing Nutrients Through Rotation

Plants require the macronutrients nitrogen (N), phosphorus (P), and potassium (K) to grow. Some crops deplete these nutrients, while others, like legumes, contribute to replenishing them. A well-designed rotation ensures these vital nutrients remain balanced.

Categorizing Crops

Heavy Feeders: Crops like corn, tomatoes, and cabbage pull significant nutrients from the soil.

Light Feeders: Vegetables such as carrots, onions, and beets require fewer nutrients.

Soil Builders: Legumes such as beans, peas, and clover fix nitrogen and improve soil fertility.

A typical four-year rotation plan might look like:

Legumes: Beans, peas, and alfalfa enrich the soil with nitrogen.

Heavy Feeders: Tomatoes, corn, and cabbage utilize the added nutrients.

Light Feeders: Carrots, onions, and beets use fewer nutrients, giving the soil a break.

Cover Crops: Clover, rye, and buckwheat replenish organic matter, reduce erosion, and suppress weeds.

By cycling through this plan, mini-farmers reduce the need for synthetic fertilizers and support long-term soil health.

Disrupting Pest and Disease Cycles

Many pests and diseases are specific to certain crops. Growing the same crop in the same space repeatedly creates ideal conditions for infestations. Rotation interrupts these cycles.

Examples:

- Tomatoes and potatoes are vulnerable to soil-borne fungal infections like fusarium wilt and verticillium.
- Cabbage and broccoli are prone to clubroot.
- Carrots and parsnips often suffer from root-knot nematodes.

Switching crop families in a field or garden bed helps reduce pest pressure naturally.

Enhancing Soil with Cover Crops

Cover crops—also called green manure—play a vital role in soil restoration between major planting cycles. Their benefits include:

- Preventing erosion
- Boosting organic matter
- Fixing nitrogen
- Suppressing weeds

Excellent cover crop options include mustard, buckwheat, winter rye, and clover. When incorporated into crop rotations, they improve long-term productivity.

Practical Rotation Tips for the Mini-Farm

- Start with a soil test to assess nutrient levels.
- Avoid following one crop with another from the same family (e.g., don't plant tomatoes after peppers).
- Divide your garden into zones, and rotate crops across those zones each year.
- Use companion planting to complement rotation benefits by enhancing plant health and repelling pests.

Succession Planting: Harvesting Year-Round

Succession planting is a smart strategy used by farmers and gardeners to maintain a steady harvest. Instead of planting everything at once, you stagger plantings, replant after harvests, and combine fast and slow growers to keep your garden productive.

Key Benefits of Succession Planting

Maximized Space Use: Keep beds full with multiple plantings per season.

Continuous Harvest: Reduce gaps in production and enjoy a steady stream of fresh vegetables and herbs.

Minimized Waste: Prevent harvest overloads and avoid spoilage.

Reduced Pest and Nutrient Strain: Rotation within succession helps with sustainability.

Efficient Use of Land: Replanting keeps soil productive throughout the growing season.

Techniques for Succession Planting

1. Staggered Planting

Plant the same crop at intervals. For instance, sowing lettuce every two weeks ensures a rolling harvest rather than one large yield.

2. Different Maturity Varieties

Use early, mid-, and late-season cultivars of the same crop to extend the harvest window.

3. Sequential Crop Replacement

Replace harvested crops with a different crop. For example, follow spring peas with bush beans in the same bed.

4. Interplanting Fast and Slow Growers

Pair quick growers like radishes with slower crops like carrots. The radishes will be harvested before the carrots mature, making room for growth.

5. Relay Planting

Begin planting the next crop before the current one finishes. For example, start tomato seedlings in the same space where spring greens are nearing the end of their cycle.

Creating Your Succession Plan

- Choose crops with different maturity timelines.
- Note seeding and harvest dates to maintain your rhythm.
- Avoid planting crops from the same family back-to-back.
- Use a calendar to manage transitions and keep planting continuous.
- Utilize season extension tools to push your growing season further.

Extending the Growing Season: Structures That Work

To maximize output beyond the natural growing window, mini-farmers use structures like cold frames, hoop houses, and greenhouses to shield crops from weather and temperature fluctuations.

Why Extend the Season?

- Enables early spring planting and late fall harvests
- Protects crops from frost, wind, and heavy rains
- Increases productivity and profitability
- Allows for more crop variety, including tender vegetables

Cold Frames: Simple and Effective

Cold frames are unheated, low-cost boxes with transparent lids that trap solar heat.

How to Build One:

- Use bricks, wood, or cinder blocks for the frame
- Cover with a clear plastic or glass lid (repurposed windows work great)
- Face the opening south for maximum sunlight
- Vent on warm days to prevent overheating

Best Crops for Cold Frames:

Leafy greens (lettuce, spinach, kale), root veggies (carrots, radishes, beets), and hardy herbs (parsley, cilantro)

Hoop Houses: Affordable and Versatile

Made of metal or PVC hoops covered with plastic sheeting, hoop houses create tunnel-like shelters.

Setup:

- Create arches over raised beds using pipes
- Drape with transparent plastic
- Secure with clamps or weights
- Vent on warm days with roll-up sides

Great for: Tomatoes, peppers, hardy greens, early root vegetables, and brassicas.

Greenhouses: Full-Scale Control

Greenhouses are enclosed structures with clear walls and roofs that offer maximum environmental control.

Types:

Cold Greenhouses: Use solar heat only; great for extending the growing season

Heated Greenhouses: Use heaters for year-round production

Hydroponic Greenhouses: Grow without soil using nutrient-rich water

Best for: Warm-weather crops (tomatoes, eggplants), herbs, microgreens, and specialty produce like turmeric or ginger.

Choosing the Right Structure

- Small-scale or tight budget? Go for cold frames or hoop houses.

- Larger farm or year-round goals? A greenhouse is worth the investment.

- Mild winters? Row covers or mini tunnels may be all you need.

Growing Smarter Year-Round

By embracing strategies like crop rotation, succession planting, and season extension, mini-farmers can cultivate more food with fewer resources. These time-tested techniques promote soil vitality, deter pests naturally, and stretch the growing season far beyond the norm. With thoughtful planning and smart structure use, even the smallest patch of land can yield abundant, healthy, and sustainable harvests all year long.

CHAPTER 10: FRUIT TREE AND VINE GROWING – PERENNIAL CROPS

Growing perennial crops—such as fruit trees and vines—is a rewarding long-term investment in food security and self-sufficiency. Unlike annuals that require yearly replanting, perennials often demand less maintenance and can produce abundantly for many years. The key to success lies in choosing varieties that align with your climate, soil, and space.

Know Your Growing Zone

Start by identifying your USDA Hardiness Zone (or equivalent in your region). This will guide you toward varieties that are well-suited to your local conditions. When selecting trees and vines, consider:

- Chill hour requirements (the number of cold hours needed for fruit production)

- Heat tolerance for warm climates

- Frost resistance to endure late spring or early autumn chills

Best Fruit Trees for Small-Scale Growing

Mini-farmers benefit most from compact and manageable varieties that deliver high yields in limited space:

- **Apples (Gala, Fuji, Honeycrisp):** Dwarf and semi-dwarf trees thrive in smaller plots.
- **Pears (Bartlett, Shinseiki):** European and Asian varieties are space-efficient and versatile.
- **Stone Fruits (Peaches, Plums, Cherries, Apricots):** Require well-draining soil and yearly pruning.
- **Citrus (Meyer Lemon, Key Lime, Tangerine):** Ideal for warm regions or container growing.
- **Figs (Brown Turkey, Celeste):** Hardy and drought-resistant, well-suited for dry climates.

Ideal Vines for Mini-Farms

Vines are excellent space-savers and can be trellised vertically:

- **Grapes (Concord, Thompson Seedless, Muscadine):** Thrive with support and full sun.
- **Kiwis:** Hardy and fuzzy varieties are available—plant both male and female for pollination.
- **Passionfruit:** A tropical-flavored vine that excels in warm climates.
- **Blackberries and Raspberries:** Thornless types are easier to manage and very productive.

Making the Most of Limited Space

Maximizing yields in tight areas is all about smart growing techniques:

- **Espalier Training:** Grow trees flat against walls or fences to conserve space.
- **Dwarf Varieties:** Deliver full-sized fruit on compact trees.
- **Container Growing:** Citrus and fig trees thrive in large pots and can be moved as needed.
- **Vertical Trellising:** Ideal for grapes, kiwis, and passionfruit, using upward space efficiently.

Pollination, Pruning, and Pest Management

Understanding Pollination Needs

Fruit production depends on successful pollination. Some trees self-pollinate, while others need a nearby partner.

- **Self-pollinating:** Peaches, apricots, sour cherries, and some apple and pear types
- **Cross-pollinating:** Many apple, pear, and plum varieties require a compatible tree nearby
- **Manual pollination:** In areas with few natural pollinators, a soft brush can be used to move pollen between flowers

Pruning for Productivity

Regular pruning is essential for healthy growth, higher yields, and disease prevention.

- **Winter pruning:** Encourages vigorous spring growth by removing dead or overcrowded branches
- **Summer pruning:** Helps shape the tree and control overgrowth
- **Thinning fruit:** Prevents limb breakage and improves fruit size and quality
- **Support structures:** Some vines and young trees benefit from staking, tying, or training along trellises

Integrated Pest Management (IPM)

Protecting fruit trees and vines naturally enhances productivity while reducing chemical use:

- **Companion planting:** Use garlic, basil, and marigolds to deter pests like aphids and mites
- **Beneficial insects:** Encourage ladybugs, lacewings, and praying mantises for natural pest control
- **Organic sprays:** Copper fungicides, insecticidal soap, and neem oil can prevent fungal diseases and pests
- **Physical barriers:** Use netting and row covers to protect fruit from birds and animals
- **Frequent monitoring:** Check leaves and fruit regularly for early signs of issues

Planning for Long-Term Growth and Sustainability

Building a Resilient Orchard

A thoughtfully designed orchard increases productivity and minimizes the impact of environmental stress.

- Test your soil for pH, nutrient levels, and drainage capacity before planting
- Apply mulch to conserve moisture, reduce weeds, and feed the soil
- Install drip irrigation to maintain consistent moisture without overwatering
- Use organic amendments such as compost and manure to enrich soil slowly and naturally
- Stick to seasonal schedules for pruning, fertilizing, and overall maintenance

Keeping Fruit Flowing Year-Round

Diversifying your perennial crops can ensure a steady supply of fresh fruit throughout the seasons:

- Stagger harvests by planting early-, mid-, and late-season varieties
- Store and preserve: Apples, pears, and persimmons store well; berries can be dried, frozen, or turned into jams
- Extend the season using greenhouses or high tunnels to protect tender crops from frost

The Financial Upside of Perennials

Fruit trees and vines offer more than food—they're a smart investment for small-scale growers:

- Lower recurring costs: Once established, perennials don't need replanting every year
- Income potential: Sell fresh fruit, jams, jellies, dried products, or even nursery plants
- Added property value: A mature orchard increases land appeal and utility

Perennial crops—like fruit trees and vines—offer smallholder farmers a sustainable and profitable path to self-sufficiency. With careful selection, proper planting techniques, seasonal care, and eco-friendly practices, even the smallest mini-farm can become a fruitful sanctuary for years to come. Growing perennials isn't just about enjoying fresh, homegrown produce—it's about investing in a resilient, rewarding, and independent lifestyle.

PART 3:
ORGANIC PEST & DISEASE MANAGEMENT

CHAPTER 11: DISEASE AND PEST CONTROL – AN INTEGRATED APPROACH

While pests and diseases are inevitable in any farming system, they don't have to derail your success. With the right knowledge and techniques, you can manage them effectively—reducing your reliance on chemical pesticides and maintaining a healthier farm environment. Early detection and prompt action are crucial to protecting your crops and keeping your mini-farm thriving.

Understanding the Pest-Disease Cycle

Pests and diseases flourish when growing conditions favor them. Poor soil health, overcrowded planting, improper watering, and lack of biodiversity create the perfect breeding ground for infestations and infections. Fungal, bacterial, and viral diseases are more likely to affect stressed or weakened plants. Recognizing the environmental factors behind pest outbreaks is the first step to managing them.

Common Pests in Small-Scale Farming

• Aphids

Tiny, soft-bodied insects that suck sap from plant tissues, causing curled leaves and stunted growth. They attract ants with their sweet honeydew excretions, which also lead to sooty mold. Commonly found on legumes, leafy greens, and fruit trees.

Control: Use neem oil, spray off with water, or introduce ladybugs.

• Caterpillars and Worms

Larvae of moths and butterflies that chew on stems and leaves, especially damaging to tomatoes, brassicas, and fruit trees.

Control: Handpick regularly, introduce parasitic wasps, and use floating row covers.

• Whiteflies

Tiny white insects that suck sap and cause leaf yellowing. They also transmit viral diseases between crops.

Control: Use reflective mulch, insecticidal soap, and encourage predators like lacewings.

• Spider Mites

Microscopic pests that feed on chlorophyll, leading to speckled leaves and fine webbing. Thrive in hot, dry climates.

Control: Raise humidity, use neem oil, and introduce predatory mites.

• Slugs and Snails

Feed on tender seedlings and leaves, leaving behind irregular holes and slimy trails. Most active in moist conditions.

Control: Handpick at night, use beer traps, or sprinkle diatomaceous earth.

• Colorado Potato Beetle

Brightly colored beetles that devour potato, tomato, and eggplant leaves. A severe infestation can strip plants bare.

Control: Handpick adults and larvae, use Bt (Bacillus thuringiensis), and cover rows.

Common Plant Diseases

• Powdery Mildew

A fungal infection that leaves a white, powdery coating on leaves. Common in cucumbers, peas, and squash.

Prevention: Improve air circulation, avoid overhead watering, and apply sulfur-based fungicides.

- **Early and Late Blight**

Fungal diseases that cause dark spots on leaves, stems, and fruit, especially in tomatoes and potatoes.

Control: Remove affected plants, rotate crops, and use copper-based fungicides.

- **Root Rot**

Usually caused by overwatering and fungal pathogens, leading to yellowing and wilting of plants.

Prevention: Improve drainage and inoculate soil with beneficial fungi like mycorrhizae.

- **Clubroot**

Affects brassicas like cabbage and broccoli, causing swollen, misshapen roots and stunted growth.

Control: Maintain soil pH near 7.2, remove infected plants, and rotate crops.

- **Fusarium and Verticillium Wilts**

Soil-borne fungi that cause yellowing and wilting in crops such as tomatoes and peppers.

Control: Choose resistant varieties, rotate crops, and solarize soil before planting.

Integrated Pest and Disease Management (IPM)

IPM is a comprehensive approach that combines multiple methods to reduce pest and disease damage while maintaining ecological balance. Here's how to build an effective strategy:

Preventive Measures

- Choose disease-resistant seeds and plant varieties.
- Practice proper spacing for good air circulation.
- Rotate crops to prevent the buildup of soil-borne pests.
- Keep weeds under control—they often harbor diseases and pests.

Biological Controls

- Introduce beneficial insects like ladybugs, parasitic wasps, and nematodes.
- Create bird- and bat-friendly habitats.
- Use companion planting, such as marigolds with tomatoes, to deter pests naturally.

Cultural Practices

- Water early in the day so leaves dry by evening.
- Remove diseased plant material immediately.
- Use mulch to retain soil moisture and suppress weeds.

Mechanical and Physical Controls

- Protect crops with row covers and insect netting.
- Use sticky traps to monitor and capture flying insects.
- Manually remove pests like beetles and caterpillars.

Natural and Organic Remedies

- Apply insecticidal soap, horticultural oil, or neem oil for soft-bodied pests.
- Use compost tea and seaweed extract to boost plant immunity.
- Apply baking soda sprays to prevent fungal diseases.

Regular Monitoring and Record-Keeping

- Inspect plants frequently for early signs of trouble.
- Track outbreaks and treatment results in a farm journal.
- Adjust strategies based on seasonal patterns and past success.

Creating a Self-Sustaining Pest Defense

Instead of depending solely on reactive measures, sustainable farming emphasizes long-term health and biodiversity to prevent outbreaks before they begin.

Companion Planting Strategies

Plant combinations that work together can enhance growth, repel pests, and attract helpful insects.

- **Marigolds + Tomatoes/Peppers –** Repel aphids, whiteflies, and nematodes.
- **Basil + Tomatoes –** Improves flavor and deters thrips and hornworms.
- **Garlic/Onion + Cabbage/Carrots –** Repels aphids, cabbage worms, and carrot flies.
- **Nasturtiums + Squash/Cucumbers –** Acts as a trap crop for aphids and squash bugs.
- **Radish + Spinach –** Draws flea beetles away.
- **Dill + Brassicas –** Attracts predators like ladybugs and lacewings.

Encouraging Beneficial Insects

Attracting natural predators can significantly reduce pest pressure.

- **Ladybugs –** Eat aphids, scale insects, and whiteflies.
- **Parasitic Wasps –** Lay eggs inside harmful insects.
- **Hoverflies –** Larvae feed on mealybugs and whiteflies.
- **Ground Beetles –** Target root maggots and slugs.

To attract them:

- Plant nectar-rich flowers like yarrow and fennel.
- Avoid synthetic insecticides.
- Create habitats with mulch beds and insect hotels.

DIY Organic Sprays and Physical Barriers

For gardeners looking for hands-on control methods, these homemade solutions are cost-effective and eco-friendly.

Neem Oil Spray

Ingredients: 1 tsp cold-pressed neem oil, 1 tsp dish soap, 1-quart warm water

Instructions: Mix in a spray bottle, shake, and apply every 7–10 days to leaves (avoid midday sun).

Garlic-Chili Spray

Ingredients: 5 minced garlic cloves, 2 chopped chili peppers, 1 quart water, 1 tsp dish soap

Instructions: Blend garlic and chili in water, strain, mix in soap, and spray plants as needed.

Baking Soda Fungicide

Ingredients: 1 tbsp baking soda, 1 tsp dish soap, 1 tsp vegetable oil, 1 gallon water

Instructions: Mix and spray weekly to prevent powdery mildew and other fungal diseases.

Other Handy Solutions

- **Diatomaceous Earth:** Dehydrates soft-bodied pests like aphids and slugs.
- **Beer Traps:** Attract and drown slugs and snails.
- **Crushed Eggshells:** Deter slugs.
- **Coffee Grounds:** Improve soil and repel ants and beetles.
- **Cinnamon Powder:** Natural antifungal and ant repellent.
- **Wood Ash:** Scatters to deter soft-bodied insects.

Managing pests and diseases on a small-scale farm requires vigilance, adaptation, and an integrated approach. By combining cultural practices, natural deterrents, beneficial insects, and organic solutions, you can protect your crops without compromising the health of your soil, plants, or ecosystem. Every challenge faced is also an opportunity to learn and refine your methods—leading to a more resilient and self-sustaining mini-farm over time.

PART 4:
RAISING SMALL LIVESTOCK FOR SELF-SUFFICIENCY

CHAPTER 12: EGG RAISING CHICKENS

Choosing the Right Breeds for Egg Production

Raising chickens for eggs is one of the most rewarding aspects of small-scale farming. However, not all chicken breeds are equal when it comes to laying. Selecting the right breed will determine not just how many eggs you get but also the size, color, and even the temperament of your flock. This chapter will help you understand the top breeds for egg production, what to consider when choosing hens, and how to create an ideal environment for maximum output.

Key Considerations When Selecting a Breed

Before bringing home your hens, keep the following in mind:

Egg Production Rate: Some breeds are prolific layers, while others lay fewer eggs but may offer other benefits like dual-purpose meat production or cold-hardiness.

Climate Suitability: Certain breeds perform better in colder climates; others are heat-tolerant.

Egg Size and Color: Do you prefer large brown eggs or a colorful assortment of blue, green, or speckled ones?

Temperament: Some chickens are docile and friendly—great for families and beginners—while others are flightier and more independent.

Space Needs: Active breeds need more room to roam, while calmer breeds can thrive in smaller coops.

Foraging Ability: If you plan to let your chickens free-range, choosing breeds that forage well will reduce your feed costs.

Top Chicken Breeds for Egg Production

Leghorn

Eggs per year: 280–320

Egg color: White

Temperament: Active, independent

Climate: Heat-tolerant

Why Choose Them? Leghorns are among the most productive layers. They thrive in both free-range and coop environments and are economical eaters.

Rhode Island Red

Eggs per year: 250–300

Egg color: Brown

Temperament: Hardy, friendly

Climate: Tolerates both hot and cold

Why Choose Them? A favorite for small-scale farms, these birds are low-maintenance and reliable egg producers.

Sussex

Eggs per year: 220–280

Egg color: Light brown

Temperament: Curious, calm, friendly

Climate: Cold-hardy

Why Choose Them? Sussex hens are adaptable and do well in confined or free-range setups. Their friendly nature makes them great for beginners.

Australorp

Eggs per year: 250–300

Egg color: Brown

Temperament: Gentle, quiet

Climate: Cold-hardy

Why Choose Them? Known for breaking egg-laying records, Australorps are perfect for new chicken keepers.

Plymouth Rock

Eggs per year: 200–280

Egg color: Brown

Temperament: Smart, calm

Climate: Cold-hardy

Why Choose Them? These classic backyard chickens are dependable layers with a sweet disposition.

Orpington

Eggs per year: 200–280

Egg color: Light brown

Temperament: Docile, affectionate

Climate: Cold-hardy

Why Choose Them? Great for families, Orpingtons are large, fluffy birds that are just as good for meat as they are for eggs.

Easter Egger

Eggs per year: 200–280

Egg color: Blue, green, pink, brown

Temperament: Friendly, curious

Climate: Adaptable to hot and cold

Why Choose Them? If you're looking to add a splash of color to your egg basket, Easter Eggers are a must-have.

Matching the Breed to Your Farm's Needs

If you live in a colder region, cold-hardy breeds like Australorps, Orpingtons, and Rhode Island Reds are ideal. For high-volume egg production, Leghorns and Rhode Island Reds are standouts. If you want a well-rounded backyard flock that's both productive and friendly, consider Sussex, Plymouth Rocks, or Easter Eggers.

Keeping Your Egg-Laying Flock Healthy and Productive

To get the most from your chickens, you need to provide a healthy environment, proper nutrition, and consistent care.

Space: Each hen needs 2–4 square feet inside the coop and 8–10 square feet of outdoor run space.

Housing: Build a predator-proof, well-ventilated coop with enough roosting space and clean nesting boxes.

Nutrition: Offer a layer feed with 16–18% protein, plus oyster shells for calcium and fresh greens for variety.

Lighting: Chickens need 14–16 hours of light daily to maintain peak laying, especially in winter.

Water: Keep water clean and available at all times.

Health Checks: Regularly inspect birds for signs of illness and parasites.

Whether your flock is confined or free-range, make sure their needs are met through proper diet, space, and enrichment.

Coop Design and Daily Care

Space Requirements

Provide at least 2–4 square feet per chicken in the coop and 8–10 square feet in the outdoor run to prevent stress and disease.

Ventilation

Install windows, vents, or mesh openings to promote airflow and control moisture. Adjustable vents are helpful year-round.

Nesting Boxes

One box per 3–4 hens is sufficient. Each should measure around 12x12 inches and be placed in a quiet, dark spot lined with straw or wood shavings.

Roosting Bars

Chickens like to perch at night. Allow 8–12 inches of space per bird on elevated wooden bars.

Predator Protection

Use hardware cloth instead of chicken wire and bury fencing 12 inches underground to prevent burrowing. Lock coop doors securely at night.

Feeding for Egg Production

Commercial Feeds

- Starter Feed (0–8 weeks): 18–20% protein
- Grower Feed (8–18 weeks): 16–18% protein
- Layer Feed (18+ weeks): 16–18% protein + added calcium
- Natural and Homemade Diet Options
- Mix grains, seeds, and protein sources like mealworms or fish meal.
- Include calcium sources like crushed oyster shells or limestone.
- Fresh greens such as weeds and kitchen scraps add nutrition and improve yolk color.

Supplements

- Provide grit to aid digestion.
- Avoid feeding onions, garlic, chocolate, or anything salty/spicy.

Establishing a Daily Care Routine

Morning

Let hens out, check feeders, refill water, and collect early eggs.

Afternoon

Check flock health, provide snacks or greens, and ensure waterers are full.

Evening

Collect eggs, lock the coop, and inspect for any predator activity or coop repairs.

Common Health Issues and Prevention

Respiratory Infections

Symptoms: Sneezing, nasal discharge, swollen eyes

Prevention: Clean, ventilated coop and quarantine new birds

Mites and Lice

Symptoms: Feather loss, pale combs, weight loss

Prevention: Dust baths, coop cleaning, and diatomaceous earth

Bumblefoot

Symptoms: Swollen footpad

Prevention: Soft bedding, smooth perches, treat minor wounds early

Egg Binding

Symptoms: Swollen abdomen, straining

Prevention: High calcium intake, hydration, stress reduction

Coccidiosis

Symptoms: Diarrhea, lethargy, fluffed feathers

Prevention: Clean bedding, medicated chick feed

Disease Prevention Tips

- Clean tools and limit visitor access.
- Quarantine new birds.
- Maintain a clean coop.
- Vaccinate where necessary.
- Do weekly health checks and act quickly on any signs of illness.

Raising productive, happy laying hens is about more than just collecting eggs—it's about creating a healthy, sustainable system that works in harmony with your land and lifestyle. With the right breeds, thoughtful care, and consistent attention, your hens will reward you with a steady supply of fresh eggs and the joy that comes from self-reliant living.

CHAPTER 13: RAISING MEAT CHICKENS

Raising chickens for meat is an excellent way to increase your farm's self-sufficiency, provide your family with high-quality protein, and potentially generate extra income. However, success in meat production depends largely on choosing the right broiler breeds and optimizing feeding routines. The right combination results in healthy birds, efficient weight gain, and flavorful meat.

Selecting the Right Broiler Breed

Not all chickens grow or convert feed into meat at the same rate. Your choice of breed affects growth speed, feed efficiency, meat quality, and how well the birds fit into your farm setup. Some fast-growing hybrids are ideal for quick turnover, while heritage breeds are valued for taste and sustainability.

1. Cornish Cross

- **Growth Rate:** Reaches 4–6 lbs in 6–8 weeks
- **Traits:** Extremely fast-growing, large breasts, high feed efficiency
- **Notes:** Needs a controlled diet and limited activity to prevent leg issues and heart strain
- **Best for:** Semi-confined setups with close monitoring

2. Freedom Rangers / Red Rangers / Rainbow Rangers

- **Growth Rate:** Ready in 10–12 weeks
- **Traits:** More active and natural growth pattern, darker meat, better foragers
- **Best for:** Pasture-raising, rotational grazing systems, or farmers seeking fuller flavor

3. Jersey Giant

- **Growth Rate:** 16–21 weeks
- **Traits:** Large, flavorful birds; slower growth but hardy
- **Best for:** Heritage meat production with an eye on sustainability

4. Bresse Chicken

- **Growth Rate:** Slow-growing but highly prized
- **Traits:** Richly marbled meat, gourmet quality
- **Notes:** Requires a specific dairy-enhanced diet for best flavor
- **Best for:** Niche markets and farmers seeking culinary excellence

5. Delaware

- **Growth Rate:** Moderate
- **Traits:** Dual-purpose, hardy, and good foragers
- **Best for:** Sustainable backyard operations with diverse needs

Feeding for Optimal Meat Production

A broiler's nutrition directly impacts how fast and how well it grows. Feed programs are usually broken down into three phases, each tailored to the bird's stage of development.

1. Phase-Based Feed Plan

Starter Feed (0–3 weeks)

- Protein: 20–24%
- High in essential amino acids for rapid skeletal and muscle growth
- Often medicated to prevent coccidiosis, but non-medicated versions are available for organic operations

Grower Feed (3–6 weeks)

- Protein: 18–20%
- Supports steady development and prevents excess fat accumulation
- Balanced with quality carbs and fats

Finisher Feed (6+ weeks)

- Protein: 16–18%
- Encourages desirable meat texture and fat marbling
- Commonly includes corn or wheat to boost flavor

2. Commercial vs. Homemade Feeds

Commercial Feeds

- Convenient and nutritionally balanced
- Available in both medicated and non-medicated forms
- More expensive but ensures consistent growth

Homemade Feeds

- Allows full control over ingredients (organic, non-GMO, etc.)
- Often includes grains, legumes, seeds, fish meal, and calcium supplements
- Requires careful formulation and regular adjustments

3. Natural Additives for Health and Growth

- **Apple Cider Vinegar:** Aids digestion and boosts immunity
- **Fermented Feed:** Enhances gut health and nutrient absorption
- **Garlic and Oregano:** Natural antimicrobials
- **Kelp Meal:** Rich in trace minerals for overall health

4. Feeding Schedule Management

- **Ad Libitum Feeding (free access):** Encourages rapid growth but must be monitored for overconsumption
- **Restricted Feeding:** Slows growth slightly, which can reduce leg and heart issues in fast-growing breeds
- **Pasture Supplementation:** Especially effective with active breeds like Red Rangers, as it reduces feed costs and enhances flavor through a natural diet of insects and plants

Raising Meat Birds Humanely and Effectively

Ethical meat production isn't just better for the chickens—it's better for your farm. Reducing stress and prioritizing welfare improves meat quality, farm sustainability, and your peace of mind.

Choosing the Right Breed

Cornish Cross offers quick returns, but Red Rangers and heritage breeds provide natural behavior and fuller taste. Choose based on your goals—speed, flavor, sustainability, or all three.

Brooding: The Critical First Weeks

New chicks require warmth, safety, and careful monitoring. For the first week, maintain temperatures around **95ºF (35ºC)**, reducing by 5ºF weekly until birds are fully feathered.

Essentials for Brooding:

- Soft bedding (like pine shavings) to absorb waste and reduce ammonia
- Proper ventilation without cold drafts
- Starter feed with 20–22% protein
- Clean, accessible water at all times
- Minimum 0.5 sq ft per chick to prevent crowding

Housing and Space Requirements

Healthy birds grow better in clean, stress-free environments. Whether indoors or pasture-based, ensure they have room to roam.

- **Indoor Housing:** At least 1.5 sq ft per bird
- **Outdoor Runs:** 5–10 sq ft per bird for pasture-based systems
- **Coop Requirements:** Dry bedding, predator-proof fencing, fresh air, and access to daylight

Nutrition Through the Growth Cycle

A staggered feed approach ensures each stage of development is supported properly:

0–3 Weeks: High-protein crumble for muscle and skeletal growth

4–6 Weeks: Moderate-protein feed to sustain development

7+ Weeks: Lower-protein finishing feed to refine meat texture and flavor

Water is equally vital. Ensure a constant supply of clean, fresh water to support digestion and regulate body temperature.

Supporting Health and Reducing Stress

Stress can negatively affect both growth and meat quality. Prioritize welfare to support efficient and humane production.

Best Practices:

- Handle birds gently and minimally
- Maintain a regular light/dark cycle
- Keep birds in consistent groups to avoid pecking order disruptions
- Use strong enclosures to deter predators

Ethical Processing and Butchering

Processing should be quick, respectful, and cause as little distress as possible.

Guidelines for Humane Slaughter:

- Keep birds calm and fed until shortly before processing
- Use mechanical or electrical stunning for unconsciousness
- Perform a swift, accurate cut to the carotid artery
- Scald and pluck efficiently, then chill and eviscerate to preserve meat quality

Troubleshooting Common Challenges

Even with good care, challenges arise. Here's how to handle a few of the most common:

1. Health Issues from Rapid Growth (Especially Cornish Cross)

- **Leg Problems:** Provide sufficient space and balanced feed
- **Heart Strain:** Use restricted feeding or slower-growing breeds

2. Disease Prevention

- Avoid overcrowding and damp conditions
- Use medicated feed or natural probiotics as needed
- Sanitize waterers and feeders regularly
- Ventilate well to prevent respiratory infections

3. Feed Costs

- Buy in bulk or grow your own ingredients
- Use spill-proof feeders
- Ferment feed to improve absorption and reduce waste

4. Predator Threats

- Install electric fencing and covered runs
- Use guardian animals like dogs or geese
- Lock coops securely at night

Raising meat chickens can be one of the most practical and fulfilling parts of small-scale farming. With thoughtful planning, ethical practices, and the right breed selection, you'll produce nutritious, homegrown meat while supporting your land's health and your family's self-sufficiency.

CHAPTER 14: BUTCHERING AND PROCESSING

A Practical and Humane Guide for Home Growers

Butchering and processing your own animals is a powerful step toward true self-sufficiency. While it can feel daunting at first, doing it ethically and effectively ensures the animals you've raised are treated with respect—and it provides your household with high-quality, home-grown meat. With preparation and care, home processing can be a fulfilling and responsible part of your mini-farming journey.

The Ethics of Home Processing

Home growers have a distinct advantage over commercial producers: the ability to offer animals a low-stress life and a calm, humane end. Unlike industrial systems where animals may suffer from transport stress or inhumane conditions, home butchering can be carried out in a peaceful, familiar environment.

Key ethical practices include:

- Reducing stress through quiet, gentle handling
- Using quick and humane slaughter techniques
- Making full use of the animal to minimize waste
- Creating a clean, efficient setup that respects the animal and the process

Preparing for Butchering

Before you begin, there are a few important steps to take:

- **Fasting:** Poultry should be fasted for 12–24 hours before processing to empty their digestive systems. This reduces mess during evisceration and helps prevent contamination.
- **Gentle Handling:** Keep animals in a calm, familiar space before slaughter to reduce stress hormones, which can also affect meat quality.

Essential Tools:

- Sharp knives
- Killing cone or restraint system
- Large pot or scalder
- Plucking tool or mechanical plucker
- Gutting tools (shears, spoons, etc.)
- A clean worktable with access to running water

Humane Slaughter Techniques

The method of slaughter varies depending on the animal. For poultry, the most common and humane approaches include:

Killing Cone Method

A cone gently restrains the bird head-down, reducing movement and keeping the bird calm. A quick, deep cut to the carotid arteries results in near-instant unconsciousness and minimal suffering.

Cervical Dislocation

Manually separating the vertebrae of the neck can be effective if done correctly—but requires skill to ensure it's fast and painless.

Captive Bolt or Electrical Stunning

For larger animals, such as pigs or goats, pre-slaughter stunning is used to render the animal unconscious before bleeding out. This prevents pain and ensures a swift end.

Plucking and Evisceration (Poultry)

Once the bird is dispatched:

1. **Scalding:** Submerge the bird in 130–150ºF water for 30–60 seconds. This loosens the feathers for easier removal.

2. **Plucking:** Remove feathers by hand or using a mechanical plucker.

3. **Evisceration:** Make a careful incision to remove internal organs. Avoid puncturing the intestines to prevent contamination.

Processing Larger Livestock

If you're raising goats, pigs, or rabbits, humane slaughter also involves:

- Stunning with an electric device or captive bolt
- A clean and swift cut to the jugular vein and carotid artery
- Bleeding out fully before skinning or dressing
- Custom butchering the carcass into usable cuts for cooking, preserving, or freezing

Meat Preservation and Storage

Proper handling after slaughter is essential for flavor, tenderness, and food safety.

- Poultry: Let rest in refrigeration for 12–24 hours before cooking or freezing.
- Red Meat: Benefits from aging for several days in refrigeration to enhance tenderness and taste.
- Vacuum Sealing or Freezer Bags: Prevent freezer burn and extend storage time.
- Curing and Smoking: Add flavor and extend shelf life, particularly for pork or specialty meats.

Sanitation and Food Safety

Keeping your setup clean and your practices consistent is key to safe, high-quality meat.

- Sanitize tools and surfaces before and after use.
- Keep meat at or below 40°F (4°C) during processing.
- Chill processed meat quickly and store it at the correct temperature.
- Compost or dispose of inedible parts responsibly to avoid pests.

Building a DIY Chicken Plucker and Processing Station

Feather plucking is one of the most time-consuming parts of poultry processing. A DIY chicken plucker can significantly cut down on labor while ensuring clean, consistent results.

DIY Chicken Plucker Materials

- 55-gallon food-grade drum (plastic or stainless steel)
- Rubber plucker fingers
- 0.5–1 HP electric motor
- Pulley and belt system
- Plywood or aluminum base
- Waterproof sealant, bolts, screws, and on/off switch

Construction Overview

1. **Prepare the Drum:** Clean and sanitize. Drill evenly spaced holes near the bottom for the rubber fingers.

2. **Insert Fingers:** Push the rubber fingers into holes tightly.

3. Build the Base and Mount Motor: Ensure the motor is secure and aligned with the drum's axle.

4. **Connect Pulley System:** Allows the motor to rotate the drum efficiently.

5. **Safety Check:** Wire the motor through a switch. Use sealant to protect against moisture.

6. **Operation:** After scalding, place the bird in the drum for 15–30 seconds to pluck. Rinse and move to evisceration.

Setting Up a Processing Station

An organized, hygienic station makes a big difference in ease and safety. Here's what you'll need:

- **Killing Cone Station:** For safe, humane dispatch
- **Scalding Tank:** 140–150°F water for feather release
- **Plucking Area:** Manual or plucker setup with a drainage basin
- **Evisceration Table:** Stainless steel or food-grade plastic surface
- **Chilling Tank:** Ice water bath to cool birds post-processing
- **Packaging Area:** Vacuum sealer or freezer bags, plus freezer or fridge

Storing and Handling Processed Meat

Once your birds are processed, follow these guidelines to keep your meat safe and fresh:

Hygiene and Tools

- Wear gloves and aprons
- Sanitize knives, tables, and tools thoroughly
- Use sharp knives for cleaner cuts and reduced bacterial risk
- Always separate raw and cooked meat

Chilling and Freezing

- Ice bath chill processed birds for 30–60 minutes
- Store refrigerated poultry at 32–40°F (0–4°C)
- Freeze at 0°F (**-18°C**) or lower

Packaging Options

- **Vacuum Sealing:** Extends shelf life up to 12 months
- **Freezer Paper:** Great for 6–9 months, moisture-resistant
- **Plastic Freezer Bags:** Squeeze out air and label with date

Thawing Safely

- **Fridge Method:** Safest—thaw over 24 hours
- **Cold Water Method:** Submerge in sealed bag under cold running water
- Avoid Room Temp Thawing: Bacteria multiply quickly

Safe Cooking

- Cook chicken to an internal temperature of 165°F (75°C)
- Refrigerate leftovers within 2 hours after cooking

Processing your own meat is not just a matter of food—it's an act of responsibility. When done thoughtfully, it connects you more deeply to your farm, your animals, and the food on your table. With the right knowledge and tools, you can provide high-quality, ethically raised meat for your family while honoring every step of the process.

PART 5:
PRESERVING AND SELLING YOUR HARVEST

CHAPTER 15: CANNING, FREEZING, AND DEHYDRATING THE HARVEST

Preserving the Harvest: Freezing, Canning, and Dehydrating for Year-Round Food Security

Preserving the harvest is an essential skill for any mini-farmer committed to self-sufficiency, reducing food waste, and enjoying seasonal abundance all year long. Whether you're looking to stock your pantry, freezer, or emergency food reserves, understanding how to properly can, freeze, or dehydrate produce allows you to lock in flavor, nutrition, and quality.

In this chapter, we'll walk through the basics and best practices of home food preservation—covering everything from safe canning techniques to long-term freezer storage and dehydrating methods.

The Foundations of Food Preservation

Preserving food isn't new—our ancestors relied on it for survival. Modern techniques have improved safety and shelf life, but the goals remain the same: keep food safe, reduce waste, and ensure year-round availability.

Three Primary Methods:

- **Canning:** Uses heat and airtight jars to kill microbes and enzymes that cause spoilage.
- **Freezing:** Slows enzyme activity and halts bacterial growth by keeping food below 0°F.
- **Dehydrating:** Removes moisture to inhibit microbial growth, extending shelf life.

Choosing the right method depends on the type of food, storage conditions, and how long you want to keep it.

Canning: Water Bath vs. Pressure Methods

Canning is one of the most reliable ways to store food without refrigeration. But not all foods are canned the same way—acid level determines the safest method.

Water Bath Canning

Ideal for **high-acid foods** (pH 4.6 or lower), such as:

- Tomatoes (with added acid)
- Pickles
- Fruit preserves and jams
- Salsa

Steps:

1. Sterilize jars and lids by boiling them for at least 10 minutes.

2. Prepare and pack food into jars, leaving proper headspace.

3. Wipe rims clean, apply lids, and tighten bands finger-tight.

4. Submerge jars in boiling water (1 inch above the tops) and process according to the recipe and altitude.

5. Let cool for 12–24 hours before checking seals and storing in a cool, dark place.

Pressure Canning

Required for low-acid foods (pH above 4.6), such as:

- Meat and poultry
- Green beans, corn, carrots
- Soups and stews

Steps:

1. Prepare the food and sterilize jars.

2. Pack jars with the proper headspace.

3. Load jars into a pressure canner with the recommended amount of water.

4. Heat until steam vents for 10 minutes, then close the vent and bring to the correct PSI.

5. Process according to the recipe and altitude, then allow the canner to depressurize fully before opening.

6. Let jars cool undisturbed for 24 hours, check seals, and store properly.

Pressure canning is vital for safety with low-acid foods, as it destroys Clostridium botulinum, the bacteria that causes botulism.

Freezing: Fast, Easy, and Reliable

Freezing is one of the simplest and most versatile ways to preserve food. Done correctly, it maintains most of the food's nutrients, flavor, and texture.

Best Practices:

- **Blanch Vegetables:** Briefly boil, then shock in ice water to stop enzymes that degrade quality.
- **Portion Smart:** Freeze in meal-size batches to avoid thawing and refreezing.
- **Use Proper Containers:** Choose freezer-safe bags or containers that prevent air and ice crystals.
- **Label Clearly:** Include date and contents to avoid mystery packages later.
- **Keep Cold:** Set your freezer to 0°F (-18°C) or lower.

Foods That Freeze Well:

- Berries (freeze flat first, then bag)
- Blanched vegetables: beans, peas, carrots, broccoli
- Meat and poultry (well-wrapped)
- Breads, baked goods, and cheese
- Butter and some dairy (milk may separate but is usable)

Foods That Don't Freeze Well:

- Raw potatoes, lettuce, cucumbers (they get soggy)
- Leafy greens (unless blanched)
- Cream-based soups or sauces (may separate when thawed)

Dehydration and Vacuum Sealing for Long-Term Storage

Dehydrating removes moisture, making food shelf-stable and lightweight—perfect for snacks, emergency kits, and saving space. When combined with vacuum sealing, dehydrated food can last for months or even years.

How to Dehydrate:

1. **Prep Food:** Wash, peel, and slice evenly.

2. **Pre-Treat (Optional):** Soak fruits like apples or bananas in lemon water to prevent browning.

3. **Arrange on Trays:** Don't overlap pieces.

4. **Set Temperature:**

   - **Herbs:** 95°F–105°F (35°C–40°C)
   - **Fruits:** 125°F–135°F (50°C–57°C)
   - **Vegetables:** 125°F–135°F (50°C–57°C)
   - **Jerky (meat):** 160°F–165°F (70°C–74°C)

5. **Drying Time:** Varies from 4–24 hours depending on humidity and food type.

6. **Cool and Store:** Let food cool before sealing in airtight containers or vacuum-sealed bags.

Good Candidates for Dehydration:

- Apples, bananas, strawberries
- Tomatoes, onions, carrots
- Herbs like basil, thyme, oregano
- Meats for jerky

Vacuum Sealing: Extending Freshness

Vacuum sealing protects food by removing oxygen—one of the main contributors to spoilage. It's a great companion to freezing or dehydrating.

Steps:

1. Place food in specially designed vacuum-seal bags.

2. Use a vacuum sealer to remove air and seal the bag.

3. Store sealed items in a pantry, fridge, or freezer depending on the item.

Vacuum-sealed foods last longer, resist freezer burn and maintain freshness far better than standard bags.

Tips for Storage, Packaging, and Safety

Sanitation Matters

- Wear gloves and aprons when handling meat or produce.
- Sanitize all tools and surfaces before and after use.
- Use sharp knives for cleaner cuts and reduced bacterial risk.
- Separate raw and cooked food during storage and handling.

Proper Chilling and Freezing

- Chill meats in an ice-water bath for 30–60 minutes post-processing.
- Keep refrigerated food between 32–40°F (0–4°C).
- Freeze below 0°F (-18°C).

Packaging Options:

- **Vacuum Sealing:** Best for long-term storage (up to 12 months)
- **Freezer Paper:** Great for 6–9 months; prevents moisture loss
- **Freezer Bags:** Squeeze out air before sealing; label clearly

Thawing Safely:

- Thaw in the refrigerator 24 hours in advance.
- For faster thawing, submerge sealed food in cold running water.
- Avoid leaving food at room temperature to prevent bacterial growth.

Mastering food preservation ensures your garden's bounty isn't limited to one season. Whether you're preparing tomato sauce in jars, freezing sweet corn, or dehydrating apple slices, each method helps stretch your food supply, reduce reliance on store-bought goods, and minimize waste.

Choose the technique that best fits the food, your preferences, and your storage space. With the right tools, planning, and a little practice, preserving the harvest becomes a joyful, nourishing part of your self-sufficient lifestyle.

MINI-FARMING FOR A SELF-SUFFICIENT LIFESTYLE

CHAPTER 16: USING FERMENTED FOODS FOR STORAGE AND HEALTH

Fermentation for Wellness, Preservation, and Year–Round Abundance

Fermentation is one of the oldest, most powerful tools for food preservation. Long before refrigeration and canning, people around the world preserved vegetables, dairy, and grains by encouraging beneficial microbes to do the work. Foods like sauerkraut, kimchi, yogurt, and kombucha have stood the test of time— not only for their ability to extend shelf life but also for their substantial health benefits.

By understanding the science and art of fermentation, small-scale farmers and home preservers can improve nutrition, prevent food waste, and build a more self-sufficient pantry.

The Science Behind Fermentation

Fermentation is a natural metabolic process in which bacteria, yeasts, or fungi break down carbohydrates into acids, gases, or alcohol. This transformation not only preserves food but also enhances flavor, digestibility, and nutritional value.

There are three main types of fermentation:

1. **Lactic Acid Fermentation:**

Used in sauerkraut, kimchi, yogurt, and pickles. Lactobacillus bacteria convert sugars into lactic acid, which naturally preserves and protects the food.

2. **Alcoholic Fermentation:**

Yeasts convert sugars into alcohol and carbon dioxide—essential in making wine, beer, and sourdough bread.

3. **Acetic Acid Fermentation**:

Alcohol is further converted into acetic acid (vinegar), creating a highly shelf-stable product with natural antimicrobial properties.

Health Benefits of Fermented Foods

Fermented foods do more than last longer—they also offer a host of wellness advantages:

1. **Improved Digestion and Gut Health**

Fermented foods are rich in probiotics—beneficial bacteria that promote healthy digestion. These live cultures can ease bloating, reduce symptoms of IBS, and support a healthy gut microbiome, which is key to overall wellness.

2. **Enhanced Nutrient Absorption**

Fermentation helps reduce antinutrients like phytates and oxalates that block mineral absorption. As a result, nutrients like iron, calcium, and zinc become more bioavailable.

3. **Stronger Immune Function**

A healthy gut supports a strong immune system. Probiotic-rich fermented foods help stimulate the production of antimicrobial peptides and support the body's ability to fight off harmful pathogens.

4. **Increased Vitamin and Antioxidant Content**

Fermentation can elevate levels of B vitamins (like B12 and folate), vitamin K2 (for heart and bone health), and antioxidants that protect cells from inflammation and oxidative stress.

5. **Mental Health Support**

Emerging research shows a strong connection between gut health and brain function. Fermented foods may help reduce symptoms of anxiety and depression by improving gut-brain axis communication.

Fermentation as a Storage Method

In addition to its nutritional benefits, fermentation is a low-energy, natural way to preserve food. It allows mini-farmers to extend the life of fresh produce without the need for freezing or complex equipment.

1. **Natural Preservation**

The acids and compounds produced during fermentation create an environment that prevents the growth of harmful bacteria and mold—without synthetic additives.

2. Long Shelf Life

Properly stored, fermented foods can last for months or even years:

- **Sauerkraut and kimchi:** Up to a year when refrigerated
- **Yogurt and kefir:** Several weeks fresh
- **Fermented vegetables:** 3–6 months in cold storage
- **Vinegar-based ferments:** Nearly indefinite shelf life

3. Energy-Efficient Storage

Unlike freezing or pressure canning, fermentation requires little to no energy after the initial preparation. Most ferments simply need a cool, dark place and occasional observation.

4. Reduced Food Waste

Fermentation turns surplus or imperfect produce into something valuable. Overripe fruits, excess cabbage, or aging milk can all be transformed into flavorful, nutrient-rich foods instead of ending up in the compost bin.

Popular Fermented Foods and How to Make Them

Here are simple instructions for some of the most accessible and beneficial fermented foods:

How to Make Sauerkraut

Ingredients:

- 1 medium head of cabbage (green or red)
- 1–2 tablespoons sea salt
- Optional: garlic, caraway seeds, or juniper berries for flavor

Steps:

1. Remove outer leaves and core, then finely shred the cabbage.
2. In a large bowl, sprinkle salt and massage the cabbage until it releases its juices.
3. Pack the cabbage tightly into a clean jar, ensuring it's submerged in its liquid.
4. Place a weight on top to keep the cabbage submerged. Cover loosely with a cloth or lid.
5. Store at room temperature (65–75°F) for 1–4 weeks. Taste regularly.
6. Once desired tanginess is reached, seal and refrigerate. Keeps up to a year.

How to Make Kimchi

Ingredients:

- 1 head Napa cabbage
- ¼ cup sea salt + 2 cups water
- 3 green onions, sliced
- 4 garlic cloves, minced
- 1-inch piece ginger, grated
- 2 tablespoons Korean red pepper flakes (gochugaru)
- 1 teaspoon sugar
- Optional: 1 tablespoon fish sauce

Steps:

1. Chop cabbage and soak in saltwater brine for 2 hours.
2. Mix ginger, garlic, gochugaru, sugar, onions, and fish sauce in a bowl.
3. Drain cabbage and mix thoroughly with the spice paste.
4. Pack into a jar, pressing down so the liquid covers the vegetables.
5. Ferment at room temperature for 3–7 days.
6. Once it tastes how you like it, refrigerate. Will keep for up to 6 months.

How to Make Yogurt

Ingredients:

- 1-quart whole milk
- 2 tablespoons plain yogurt with live cultures

Steps:

1. Heat milk to 180°F to kill unwanted bacteria.
2. Cool to 110°F, then stir in yogurt.
3. Pour into jars and keep warm (110°F) for 6–12 hours using a yogurt maker, oven light, or insulated cooler.
4. Refrigerate when set. Use within 1–2 weeks.

How to Make Pickles (Lacto-Fermented)

Ingredients:

- 4–5 small cucumbers
- 2 cups water
- 1 tablespoon salt
- 2 garlic cloves
- 1 teaspoon dill seeds
- 1 teaspoon mustard seeds

Steps:

1. Clean cucumbers and cut if needed.
2. Boil water with salt to make brine, let cool.
3. Pack cucumbers, garlic, and spices into a jar.
4. Pour brine to submerge fully.
5. Cover loosely and ferment for 3–7 days at room temp.
6. Once soured, refrigerate for up to 6 months.

Curing and Smoking Meats

Fermentation isn't the only traditional method that adds flavor and shelf life. Curing and smoking meats are time-tested techniques for preserving protein without refrigeration.

How to Cure Meat

Ingredients:

- 5 lbs meat (pork, beef, or fish)
- 1 cup sea salt
- ½ cup sugar
- 1 tbsp black pepper
- Optional: 1 tsp pink curing salt (for safety)

Steps:

1. Trim excess fat and pat meat dry.
2. Rub salt and sugar mixture over the meat.
3. Refrigerate in a sealed container for 5–7 days.
4. Rinse, then air-dry for 24 hours before smoking or slicing.

How to Smoke Meat

Equipment:

- Smoker (charcoal, wood, or electric)
- Wood chips (e.g., apple, cherry, hickory)

Steps:

- Preheat smoker to 225°F.
- Apply rub or marinade to cured meat.
- Smoke until internal temperature is safe:
- 165°F for poultry
- 145°F for pork and beef

Let rest before slicing. Store in fridge or freeze for long-term use.

From fermenting cabbage to curing pork, these traditional techniques connect us to the rhythms of food and the wisdom of generations. Fermentation offers a reliable, low-energy way to preserve harvests while delivering incredible nutritional benefits. Combined with meat curing and smoking, it opens the door to a more sustainable, resilient kitchen.

When you master these methods, you're not just saving food—you're preserving health, history, and the joy of truly nourishing meals.

MINI-FARMING FOR A SELF-SUFFICIENT LIFESTYLE

CHAPTER 17: SMALL-SCALE GRAIN PROCESSING

Grain farming is often associated with vast commercial fields, but small-scale growers can absolutely raise grains on a modest plot of land. Growing your own grains adds an invaluable layer of food security, self-sufficiency, and variety to your mini-farm. With a little planning and the right tools, your homestead can supply flour for baking, feed for livestock, and even ingredients for brewing—all from your backyard.

Choosing the Right Grains for Small Plots

Not all grains are suited for small-scale growing. The best grains for homesteads are those that offer high yield, minimal maintenance, and multiple uses. Here are some top choices:

- **Wheat –** Ideal for bread and baking; prefers well-drained soil in temperate climates.
- **Barley –** Versatile for food, feed, and brewing; grows well in cooler weather.
- **Oats –** Easy to grow, nutritious, and useful for both human consumption and animal feed.
- **Rye –** Hardy and tolerant of poor soil; great for cold climates.
- **Corn (Maize) –** Requires warmth and space but offers high yields for food and feed.
- **Millet –** Quick-growing and drought-resistant; great for chicken feed.

- **Quinoa –** A pseudo-grain packed with nutrients, adaptable to a variety of climates.

Soil Preparation and Planting Techniques

Healthy soil is the foundation of successful grain farming.

Soil Preparation:

1. **Choose the Right Location –** Full sun and well-drained soil are ideal.
2. **Test and Amend Soil –** Aim for a pH between 6.0 and 7.5; amend with compost or aged manure to boost fertility.
3. **Clear and Loosen –** Till or manually break up compacted soil and remove weeds.
4. **Add Organic Matter –** Improve structure and nutrient content with compost or green manure.

Planting Methods:

1. **Broadcast Seeding –** Scatter seeds evenly, then rake lightly into the soil.
2. **Row Planting –** Sow in straight rows: wheat/barley (6–8 inches apart), corn (12–18 inches apart).
3. **Watering and Mulching –** Keep soil moist during germination; mulch helps retain moisture and suppress weeds.

Maintaining Your Grain Crop

Once planted, grains require relatively low maintenance but benefit from early attention to weeding, watering, and pest control.

Weed Control:

- **Mulching –** Use straw or organic mulch to suppress weed growth.
- **Hand Weeding –** Practical for small plots and sensitive crops.
- **Crop Rotation –** Rotating with legumes helps restore soil and reduce weed pressure.

Watering Tips:

Most grains are drought-tolerant but need consistent moisture during germination and early growth.

Pest and Disease Management:

- **Aphids and Mites –** Control with insecticidal soap or introduce beneficial insects like ladybugs.
- **Rust and Blight –** Prevent fungal diseases with good air circulation and crop rotation.
- **Birds and Rodents –** Use netting, scare devices, or attract natural predators to protect your crops.

Harvesting and Processing Grains

Grains are ready to harvest when:

- Seed heads are dry and brown
- Kernels are hard and cannot be dented with a fingernail
- Stalks begin to yellow

Hand Harvesting:

Use a sickle or scythe to cut stalks near the ground. Gather into bundles and allow to dry for several days.

Threshing and Winnowing:

- Threshing – Beat or rub the seed heads to release grains from the husks.
- Winnowing – Toss the grain in front of a breeze or fan to separate chaff from grain.

Storage Tips for Long-Term Grain Viability

Proper storage is crucial to preserve flavor and prevent spoilage.

- Drying – Ensure moisture content is below 12% before storing.
- Containers – Use airtight bins, glass jars, or Mylar bags with oxygen absorbers.
- Environment – Store in a cool, dry space below 60°F (15°C).
- Rodent Protection – Choose sealed, pest-proof containers and elevate bins off the ground.

DIY Grain Processing Tools and Public Domain Thresher Designs

Processing your grain doesn't require industrial equipment. With creativity and access to public domain plans, you can build or adapt tools that make processing efficient and cost-effective.

1. Pedal-Powered Threshers

- Powered by human pedaling, these machines use a rotating drum to separate grain.
- Built with a bike frame, metal drum, and beater bars, they're effective and energy-free.

2. Hand-Cranked Threshers

- Similar in design to pedal-powered models but manually cranked.
- Ideal for processing small batches without fuel or electricity.

3. Traditional Flail Method

- A time-tested tool made of two sticks joined by a flexible joint.
- Lay stalks on a tarp or concrete and beat with the flail to release grain.

4. DIY Fan Winnowers

- Simple setup using a box or funnel and a household fan.
- Pour threshed grain from a height; chaff is blown away, leaving clean grain behind.

Storing and Grinding Grains at Home

Once grains are threshed and cleaned, you'll likely want to mill them for flour or meal.

Best Storage Methods:

- **Airtight Containers –** Use food-grade buckets with gamma-seal lids.
- **Mylar Bags –** Include oxygen absorbers for long-term pantry storage.
- **Cool, Dry Environment –** Ideal storage conditions prevent mold and pests.

Grinding Options:

1. Manual Grain Mills

Hand-cranked, electricity-free, and great for occasional use.

Models include:

- Wonder Junior Deluxe
- Country Living Grain Mill

2. Electric Grain Mills

Faster and more efficient for large batches.

Popular choices:

- Mockmill 100
- NutriMill Classic
- WonderMill

3. Stone vs. Metal Burrs:

- **Stone Mills:** Preserve nutrients and produce fine flour.
- **Metal Burrs:** Handle larger volumes but may heat the grain more.

4. DIY Grinding Options:

- **Blender or Food Processor –** For small batches of coarse flour.
- **Mortar and Pestle –** Traditional but labor-intensive.
- **Rolling Pin + Ziplock Bag –** A basic way to crush grains when nothing else is available.

Growing and processing your own grains may seem old-fashioned, but it's a profoundly empowering way to nourish yourself, reduce dependence on store-bought staples, and complete the circle of self-reliant food production. With a modest plot and simple tools, you can transform humble seeds into fresh flour, animal feed, and shelf-stable reserves.

By applying traditional knowledge and modern ingenuity, even a small homestead can produce a meaningful harvest of homegrown grains—connecting you to your land, your food, and a time-honored way of life.

PART 6:
MAKING MINI-FARMING PROFITABLE

MINI-FARMING FOR A SELF-SUFFICIENT LIFESTYLE

CHAPTER 18: MARKETING YOUR GOODS & SERVICES

Turning Your Farm into a Business: Selling What You Grow

Starting a small-scale farm business can be one of the most fulfilling ways to turn your passion for growing food into a sustainable income. Whether you're selling eggs, heirloom tomatoes, herbal teas, or artisan jams, success comes from a thoughtful balance of planning, smart marketing, and strong customer relationships. This chapter walks you through how to launch your farm venture, promote your goods, and build a loyal customer base.

Starting a Small-Scale Farm Business

1. Define Your Business Model

Before you dive into sales, take time to decide what kind of farm business you want to operate. Your model will shape everything from legal requirements to marketing strategy.

Here are a few common small-farm business types:

- **Market Farming:** Selling fresh produce through farmers' markets or CSA (Community Supported Agriculture) subscriptions.
- **Value-Added Products:** Creating items like jams, pickles, breads, soaps, or herbal teas using ingredients you grow.
- **Livestock & Animal Products:** Selling eggs, milk, cheese, honey, or pasture-raised meats.
- **Specialty Crops:** Growing mushrooms, medicinal herbs, heirloom vegetables, or microgreens for niche markets.

2. Legal and Regulatory Considerations

Running a farm business comes with important legal responsibilities:

- **Business Structure:** Choose and register a business type—LLC, sole proprietorship, or cooperative.
- **Licensing & Permits:** Research local and state requirements for selling raw or processed foods.
- **Food Safety Compliance:** Follow USDA or FDA guidelines for processing, labeling, and handling.
- **Tax Obligations:** Understand your obligations for sales tax, income tax, and any applicable agricultural exemptions.

3. Write a Business Plan

A well-crafted business plan acts as a roadmap for your farm's growth and success.

Include the following sections:

- **Mission Statement:** What is your farm's purpose or vision?
- **Market Research:** Who are your customers? What are their needs?
- **Production Plan:** What will you grow or produce—and when?
- **Financial Projections:** Estimate revenue, expenses, and profitability.
- **Marketing Strategy:** How will you attract and retain customers?

Finding Your Sales Channels

There's no one-size-fits-all approach to selling farm goods. Choose the mix of sales channels that best fits your product, personality, and customer base.

1. Farmers' Markets

Farmers' markets are an excellent way to connect directly with your community and build trust with local customers.

Tips for success:

- **Choose the Right Market:** Visit several markets to find one with good foot traffic and a customer base that matches your offerings.
- **Create an Inviting Booth:** Use clean tablecloths, clear signage, and baskets or crates to display your goods attractively.
- **Engage Visitors:** Offer samples, share recipes, and be open to questions—your story matters.
- **Track What Works:** Keep notes on popular items, best-selling days, and customer feedback.

2. Community Supported Agriculture (CSA)

CSA programs allow customers to subscribe to a regular share of your farm's harvest, creating predictable income for you and consistent value for them.

Steps to get started:

- **Set Your Share Options:** Offer flexible sizes (e.g., full or half shares) and pricing to accommodate different households.
- **Plan a Distribution System:** Choose drop-off points or offer home delivery if feasible.
- **Communicate Often:** Send weekly emails or newsletters with recipes, harvest updates, and notes about growing conditions.
- **Be Transparent:** Explain seasonal variability and potential crop failures upfront.

3. Online Sales

The internet has opened new doors for direct-to-consumer farm sales.

Ways to market online:

- **Create a Website:** Use platforms like Shopify, Wix, or WordPress to build a user-friendly store.
- **Leverage Social Media:** Share behind-the-scenes stories, planting updates, and product spotlights on Instagram, Facebook, or TikTok.
- **Use Third-Party Platforms:** Sell on Etsy, LocalHarvest, or regional farm marketplaces.
- **Offer Subscription Boxes or Meal Kits:** Curated farm bundles make great gifts and regular sales opportunities.

Smart Pricing and Building Customer Loyalty

1. Price with Purpose

Your pricing should reflect the real costs of production, labor, and value—without pricing yourself out of the market.

Key pricing tips:

- **Know Your Costs:** Factor in seeds, soil amendments, packaging, labor, transportation, and time.
- **Do Market Research:** Compare your pricing with local stores, CSAs, and other farmers' markets.
- **Highlight Your Value:** Organic certification, heirloom varieties, or regenerative practices can justify a premium.
- **Offer Discounts Strategically:** Provide bulk deals, early-bird CSA discounts, or bundled pricing to encourage larger purchases.

2. Build a Brand and Connect with Customers

Loyal customers don't just buy your products—they support your story.

Engagement strategies:

- Consistent Branding: Use a recognizable logo, packaging style, and tone across all touchpoints.
- Content Marketing: Share tips, recipes, farm updates, and customer testimonials on your blog or social channels.
- Events & Experiences: Host farm tours, canning workshops, or cooking demos to deepen the connection.
- Loyalty Programs: Offer referral bonuses, discounts for returning customers, or early access to new products.

Scaling Your Farm Business

Once you've established your foundation, look for opportunities to expand your reach:

- **Wholesale:** Partner with local restaurants, co-ops, or grocers who value fresh, local ingredients.

- **Value-Added Products:** Develop shelf-stable offerings like dried herbs, teas, sauces, or preserved goods.

- **Agri-Tourism:** Offer experiences like U-pick days, farm stays, or seasonal events to bring people to your land.

- **Collaborations:** Team up with local artisans, chefs, or herbalists to create unique products or host shared events.

Building a profitable small-scale farm business isn't just about growing great food—it's about sharing your story, engaging your community, and creating something that reflects your values. Whether you're at a bustling farmers' market, sending out CSA shares, or shipping homemade elderberry syrup from your website, your farm has the power to nourish others and sustain itself.

With thoughtful planning, authentic marketing, and a customer-first mindset, your small farm can thrive not only as a business—but as a fulfilling way of life.

CHAPTER 19: VALUE-ADDED AGRICULTURAL GOODS

Elevating Your Farm with Jams, Pickles, Baked Goods, and More

Value-added products can dramatically increase the profitability of your small farm while giving you new ways to connect with customers and express your creativity. By transforming fresh produce into jams, dried herbs, pickles, or baked goods, you not only reduce waste—you create products with longer shelf lives, higher price points, and broader market appeal.

Whether you're selling at the farmers' market, through CSA shares, or online, adding value to what you grow helps build a resilient, diversified farm business. That said, it's essential to understand the rules around food production and labeling, and to develop a brand that stands out.

Getting Started with Value-Added Products

Jams and Fruit Preserves

Jams, jellies, and preserves are among the most popular value-added items. Customers love their handmade charm, bold flavors, and long shelf life. If you grow berries, apples, plums, or rhubarb, you already have the raw ingredients for a profitable product.

Steps for Making Jams and Preserves:

1. Select high-quality fruit—fresh, ripe, and free of blemishes.

2. Wash, peel, pit, and chop as needed.

3. Cook with sugar and pectin to achieve the desired consistency.

4. Sterilize jars thoroughly to prevent contamination.

5. Use water bath canning to safely seal and preserve.

6. Label clearly with ingredients, production date, and allergens.

Drying Herbs for Profit

Drying herbs is a low-cost, high-value way to use surplus harvests. Popular options include basil, oregano, rosemary, thyme, sage, and mint. Dried herbs are ideal for culinary use, teas, and wellness products.

Drying Methods:

1. Air-Drying: Bundle herbs and hang upside down in a warm, dry, ventilated space for 1–2 weeks.

2. Dehydrator: Use a food dehydrator at 90–110°F for several hours to speed up and control the drying process.

3. Oven-Drying: Set your oven to its lowest temperature and dry herbs on trays with the door slightly ajar.

4. Store dried herbs in airtight containers away from sunlight to retain aroma and potency.

Pickles and Fermented Goods

Pickling is another delicious way to add value to vegetables like cucumbers, carrots, green beans, radishes, and onions. Fermented items like kimchi and sauerkraut are increasingly popular for their health benefits and bold flavor.

Basic Pickling Process:

1. Wash and chop vegetables.
2. Prepare a brine of vinegar, water, salt, and optional spices.
3. Pack vegetables into sterilized jars.
4. Cover with brine, ensuring full submersion.
5. For quick pickles, refrigerate. For shelf-stable products, process in a water bath.

For fermented pickles, no vinegar is needed—just saltwater brine, time, and the right storage conditions.

Baked Goods from the Farm

Farm-fresh eggs, milk, herbs, fruit, and honey are ideal for crafting wholesome baked goods. Think rustic loaves, muffins, granola bars, and fruit-filled pastries.

Tips for Success:

1. Use local, natural ingredients and highlight them in your marketing.
2. Experiment with seasonal flavors and signature recipes that reflect your farm's character.
3. Follow local food safety rules for preparation, packaging, and labeling.
4. Sell at markets, in CSA boxes, or offer pre-orders online for weekly pickups.

Understanding Food Laws and Regulations

Cottage Food Laws

Many states have Cottage Food Laws that allow small-scale producers to sell specific homemade goods without using a commercial kitchen. These laws vary by region but often include:

- **Permitted Foods:** Jams, jellies, dried herbs, baked goods, and other non-perishable items.
- **Sales Channels:** Direct sales at markets, from your farm, or online (local only).
- **Labeling Requirements:** Must include ingredient list, allergen information, farm name and contact info, and often a disclaimer like "Made in a home kitchen not inspected by the health department."
- **Income Limits:** Some states cap how much you can earn under cottage food rules.

Additional Legal Requirements

- **Food Handler Certification:** May be required depending on your location and product type.
- **Product Licensing:** Fermented or high-risk products may require extra permits or inspections.
- **Business Registration:** Even small businesses often need to register and report earnings for tax purposes.

Tip: Always check with your local health department or cooperative extension for current laws and guidelines.

Branding and Marketing Your Value-Added Goods

Build a Memorable Brand

A strong brand tells your story and helps your products stand out. Great branding inspires trust, loyalty, and word-of-mouth buzz.

Essentials of Branding:

- **Name & Logo:** Choose something clear, memorable, and connected to your farm's identity.
- **Packaging:** Use clean, attractive, eco-conscious packaging with clear, honest labels.
- **Unique Selling Proposition (USP):** Emphasize what makes your product different—local honey, heirloom fruit, herbal infusions, or organic ingredients.

Smart Sales Strategies

At the Farmers' Market

- Design an eye-catching booth with clear signage and inviting displays.
- Offer samples (if allowed) and invite customers to share feedback.
- Build relationships by remembering names and asking about favorite products.

Online and Social Media

- Create an online store using platforms like Shopify, Etsy, or Square.
- Use Instagram, Facebook, or TikTok to share your process, recipes, or "day in the life" content.
- Send out email newsletters with new product updates, seasonal features, and CSA availability.

CSA Add-Ons

- Include small-batch jams, teas, spice blends, or baked goods as optional extras in CSA shares.
- Offer special holiday bundles, sampler boxes, or gift-ready products.

Growing Your Value-Added Business

Once your product line is established, consider expanding through:

- **Wholesale accounts** with local cafés, boutiques, or grocers.
- **Specialty items** like infused oils, spice blends, syrup kits, or seasonal gift baskets.
- **Workshops** on jam-making, herb drying, or baking at home.
- **Collaborations** with local artisans, chefs, or herbalists for co-branded products.

Value-added goods are one of the most rewarding ways to make the most of your harvest. They allow you to preserve your crops, generate more income per pound of produce, and create a deeper connection with your community. Whether you're boiling jam in small batches, baking rosemary shortbread from your own herb garden, or blending tea from farm-grown mint, each jar or loaf becomes a story your customers can savor.

With attention to food safety, creative branding, and consistent quality, your homemade goods can become beloved staples in your region—and a cornerstone of your farm's long-term sustainability.

PART 7:
BUILDING A SUSTAINABLE MINI-FARM LIFESTYLE

CHAPTER 20: LONG-TERM SUCCESS AND SELF-SUFFICIENCY

Mini-farming is more than a way to grow food—it's a philosophy, a lifestyle, and a long-term commitment to living in harmony with the land. While the short-term goal might be putting fresh food on your table, the bigger picture is about reducing reliance on external systems, building resilience, and passing this knowledge down through generations. In this final chapter, we explore how to build an ecologically self-sustaining farm, expand into homesteading skills, and become a teacher and role model for your family and community.

Building a Self-Sustaining Mini-Farm

True sustainability is achieved when your farm can thrive with minimal external inputs. This means working with natural systems, reducing waste, and cultivating a diverse, balanced ecosystem.

1. Start with Healthy, Living Soil

Soil is the foundation of everything you grow. Without rich, biologically active soil, crop yields will decline, and disease pressure will increase.

- Feed the soil regularly with compost, mulch, and well-aged manure.
- Rotate crops to prevent nutrient depletion and reduce pest buildup.
- Grow cover crops, especially legumes, to fix nitrogen and build structure.
- Avoid tilling when possible, to protect soil microbes and reduce erosion.

2. Integrate Livestock with Crops

Animals and plants work best together in a closed-loop system.

- Chickens control pests and fertilize the soil while supplying eggs and meat.
- Bees pollinate fruit trees and garden crops—and provide honey and beeswax.
- Goats and sheep offer milk, meat, and fiber while grazing weeds and enriching the land.
- Ducks and geese can patrol for slugs and insects in garden areas.

When managed wisely, animals contribute fertility, pest control, food, and even income.

3. Conserve and Manage Water Efficiently

Water is a precious resource, and sustainability depends on using it wisely.

- Harvest rainwater using barrels, tanks, or cisterns.
- Install drip irrigation or soaker hoses to reduce evaporation.
- Reuse greywater from sinks or laundry for watering fruit trees and perennial beds.
- Mulch generously to retain soil moisture and suppress weeds.

4. Grow More Perennial Crops

Perennials offer long-term food security with less labor and fewer inputs than annual crops.

- Fruit trees (apples, plums, pears, walnuts) produce for decades once established.
- Berry bushes and grapevines provide abundant, annual harvests.
- Herbs like rosemary, thyme, mint, and lavender are resilient and multipurpose—great for food, medicine, and value-added goods.

The more you can rely on perennials, the more stable your food system becomes over time.

Expanding into Homesteading Skills

Mini-farming naturally lends itself to other self-reliant practices. By learning traditional homesteading skills, you reduce dependence on store-bought items and gain the satisfaction of making what you need from scratch.

1. Soapmaking and Natural Skincare

Homemade soap and skincare products are both practical and profitable. They allow you to control ingredients and eliminate synthetic chemicals from your home.

- Use ingredients from your farm—goat milk, beeswax, honey, and herbs.
- Try simple methods like melt-and-pour or explore cold-process soapmaking.
- Create lotions, salves, and balms using infused oils, dried flowers, and natural essential oils.
- Sell or gift them at markets, as part of CSA boxes, or in local shops.

2. Off-Grid and Alternative Energy

Producing your own energy enhances self-reliance and keeps your farm running in emergencies.

- Solar panels can power lights, refrigeration, and water pumps.
- Wind turbines are ideal for properties with consistent airflow.
- Wood stoves and rocket mass heaters offer efficient, fuel-independent heat.
- Biomass cookers and solar ovens provide energy-free cooking options.

Even a small system can dramatically reduce your utility bills and environmental impact.

3. Preserving Food for Year-Round Use

Nothing says self-sufficiency like a well-stocked pantry. Learning preservation techniques ensures that nothing from your harvest goes to waste.

- Canning fruits, vegetables, sauces, and meats for long-term shelf stability.
- Fermenting vegetables into sauerkraut, kimchi, and pickles—great for gut health.
- Dehydrating herbs, fruits, and jerky for compact, shelf-stable storage.
- Freezing fresh produce, broth, and baked goods when you have surplus.

Preserving food reduces dependence on stores, cuts waste and provides peace of mind during uncertain times.

Passing Down Knowledge: Family and Community Education

Self-sufficiency doesn't thrive in isolation. Teaching others ensures that the skills and values of mini-farming continue to grow, especially in a world that increasingly needs them.

1. Involving the Family

Make mini-farming a shared way of life by encouraging family members—especially children—to participate in meaningful ways.

- Assign age-appropriate chores like watering, feeding animals, or gathering eggs.
- Cook together using homegrown ingredients and preserve produce as a family activity.
- Teach basic skills like sewing, soapmaking, andcomposting.
- Encourage curiosity, responsibility, and connection to nature through daily participation.

2. Creating Community Connections

Your farm can be a hub for learning and inspiration in your local area.

- Host farm tours or open houses so others can see how a mini-farm operates.
- Offer hands-on classes on gardening, composting, or food preservation.
- Form a skill-sharing group for soapmaking, herbal remedies, beekeeping, or other homesteading crafts.
- Start a seed swap, community compost project, or backyard gardening initiative.

Sharing knowledge builds resilience far beyond your own fence line.

3. Supporting Local Food and Education Initiatives

The more people who understand the value of local food and self-sufficiency, the stronger your community becomes.

- Join or organize a farmers' market where local growers and makers can sell directly.
- Support community gardens that offer space and education for families to grow their own food.
- Partner with schools or youth programs to introduce children to growing, harvesting, and cooking fresh food.

Your farm can serve as both a provider and a teacher—helping others plant the seeds of change in their own lives.

Mini-farming is more than a method—it's a movement. A way of reconnecting with the land, restoring balance in our lives, and reclaiming the ability to provide for ourselves and others. It's about building a life rooted in intention, resilience, and the joy of creating something with your own two hands.

As you continue to grow, preserve, share, and teach, remember that your work is laying the foundation for a stronger future. The skills you develop today can nourish your family, empower your neighbors, and inspire future generations.

Your journey toward long-term self-sufficiency begins one plant, one harvest, one choice at a time—and it's a path worth walking.

CONCLUSION: YOUR PATH TO SELF-SUFFICIENCY

If you've made it this far, congratulations. You're no longer just thinking about mini-farming—you're walking the path toward a life rooted in sustainability, resilience, and purpose. This book has taken you from the basics of soil care and planting to animal husbandry, food preservation, renewable energy, and even community outreach.

Now, it's time to take what you've learned and start building the lifestyle you've always envisioned—one that's rich in connection, full of flavor, and guided by your own hands.

Core Lessons to Carry with You

- **Start Small, Grow Smart**

You don't need acres to begin. A balcony herb garden or backyard coop is enough. Build confidence through action, and let your setup expand naturally as your knowledge grows.

- **Healthy Soil Is Everything**

Great crops start with great soil. Prioritize composting, crop rotation, cover crops, and organic inputs to build long-term fertility.

- **Grow What You Actually Eat**

Focusing on foods and animals that meet your household needs reduces waste and maximizes utility.

- **Diversify Your Farm**

Blend crops, livestock, and multiple preservation methods to ensure food security, adaptability, and a healthy income stream.

- **Preserve the Surplus**

Don't let anything go to waste—learn to can, ferment, dehydrate, and freeze to enjoy your harvest year-round.

- **Make Sustainability the Goal**

A self-reliant mini-farm reuses, recycles, and regenerates. Rainwater collection, composting, perennial crops, and integrated animal systems make your farm circular.

- **Stay Curious, Stay Connected**

The journey doesn't end here. Join local groups, attend workshops, and keep learning. Every season teaches something new.

- **Most of All—Just Begin**

You'll never feel completely ready, and that's okay. Every mistake is a lesson. Every tomato, every egg, every homemade jar of jam is a step closer to self-sufficiency.

Suggested Resources for Lifelong Learning

Books

- ☑ The Resilient Gardener – Carol Deppe
- ☑ Mini Farming: Self-Sufficiency on ¼ Acre – Brett Markham
- ☑ The New Organic Grower – Eliot Coleman
- ☑ Storey's Guide to Raising Chickens – Gail Damerow
- ☑ Ball Complete Book of Home Preserving

Websites & Forums

- ☑ ATTRA (National Sustainable Agriculture Information Service) – attra.ncat.org
- ☑ Permies – www.permies.com
- ☑ Mother Earth News – www.motherearthnews.com
- ☑ Homesteading Today Forum – www.homesteadingtoday.com

Programs & Organizations

- ☑ Local Agricultural Extension Offices – Often offer free or low-cost workshops.
- ☑ SARE (Sustainable Agriculture Research & Education) – Provides grants and resources.
- ☑ CSA & Farmers' Market Networks – Great for learning from others and sharing your goods.
- ☑ Online Courses – Explore offerings from Coursera, Udemy, or universities on permaculture, organic farming, and homesteading.

Your Next Steps: Start Today

1. **Assess Your Space**

   From a window box to a quarter acre, know what you've got and how it can be used.

2. **Set a Simple Goal**

   Pick one or two starter goals—growing salad greens, raising chickens, starting a compost bin.

3. **Plant Something Today**

   Even if it's a single pot of basil. Action builds momentum.

4. **Sketch a Simple Farm Plan**

   Map out raised beds, animal shelters, compost areas, and water sources.

5. **Learn One New Skill Per Month**

   Canning, worm composting, soap-making—set a rhythm of learning.

6. **Find a Mentor or Join a Group**

   The journey is richer (and easier) when shared.

7. **Set 1-, 5-, and 10-Year Goals**

   From selling eggs to going fully off-grid, dream big and plan in steps.

8. **Track Your Progress**

   Keep a farm journal. Document what works, what doesn't, and what you've learned.

9. **Celebrate Small Wins**

   Your first ripe tomato, your first dozen eggs, your first successful harvest—pause and enjoy it.

Self-Sufficiency Is a Lifestyle, Not a Destination

Mini-farming isn't just about saving money or growing food—it's about reclaiming agency in your life. It's about nurturing your family, your community, and the land itself. Whether you go all in or take small steps, every move toward self-reliance brings deeper meaning, security, and fulfillment.

This is your beginning. Keep planting, keep learning, and enjoy every moment of the journey. The mini-farming life is rich with rewards—and your future is growing right beneath your feet.

APPENDIX

1. Planting & Harvesting Calendar (Temperate Zones)

Season	What to Plant
Early Spring	Lettuce, spinach, peas, radish, kale, onions
Mid-Spring	Potatoes, carrots, beets, broccoli, cauliflower
Late Spring	Tomatoes, peppers, beans, squash, cucumbers
Early Summer	Beans, cucumbers, zucchini, melons
Mid-Summer	Fall crops like kale, beets, carrots
Late Summer	Garlic, onions, cover crops for soil health
Fall	Spinach, radish, turnips, overwintering crops
Winter	Soil prep, composting, seed ordering, planning

2. Soil Amendment Cheat Sheet

Amendment	Benefit
Compost	Improves texture, fertility, microbial life
Aged Manure	Adds nutrients and organic matter
Leaf Mold	Enhances water retention and aeration
Bone Meal	Phosphorus-rich; supports root growth
Blood Meal	High in nitrogen for leafy crops
Wood Ash	Raises pH, adds potassium
Perlite	Improves drainage in heavy soils
Vermiculite	Boosts water retention

3. Organic Pest Control at a Glance

Pest Control	Purpose
Marigolds	Repel aphids, nematodes, whiteflies
Basil & Garlic	Deter mosquitoes, aphids, Japanese beetles
Neem Oil Spray	Controls aphids, mites, caterpillars
Diatomaceous Earth	Kills soft-bodied insects on contact
Soap Spray	Suffocates aphids and spider mites
Ladybugs & Lacewings	Natural predators of many pests

4. Food Preservation Quick Reference

Method	Best For	Tips
Water Bath Canning	Jams, jellies, tomatoes, pickles	Only for high-acid foods
Pressure Canning	Meats, beans, low-acid vegetables	Must use for safety
Freezing	Greens, berries, meat, bread	Blanch veggies before freezing
Dehydrating	Herbs, fruits, jerky	Use vacuum sealing for long shelf life
Fermenting	Sauerkraut, kimchi, yogurt, pickles	Store in cool, dark, airtight place

5. Recommended Reading & Resources

1. The Resilient Gardener by Carol Deppe
2. Seed to Seed by Suzanne Ashworth
3. The Art of Fermentation by Sandor Katz
4. The New Organic Grower by Eliot Coleman
5. USDA National Organic Program – www.ams.usda.gov

IMAGES USED UNDER LICENSE FROM SHUTTERSTOCK.COM

www.ingramcontent.com/pod-product-compliance
Lightning Source LLC
Chambersburg PA
CBHW081000120626
46546CB00010B/2975